Quick Guideline for Computational Drug Design

Authors

Sheikh Arslan Sehgal

Department of Biosciences, COMSATS Institute of Information Technology,
Sahiwal, Pakistan
State Key Laboratory of Membrane Biology,
Institute of Zoology; Chinese Academy of Sciences, Beijing, China
University of Chinese Academy of Sciences, Beijing, China

Rana Adnan Tahir

Department of Biosciences, COMSATS Institute of Information Technology,
Sahiwal, Pakistan
Beijing Key Laboratory of Separation and Analysis in
Biomedical and Pharmaceuticals, Department of Biomedical Engineering,
School of Life Sciences, Beijing Institute of Technology, China

A. Hammad Mirza

University of Chinese Academy of Sciences, Beijing, China
National Laboratory of Biomacromolecules, Institute of Biophysics,
Chinese Academy of Sciences, Beijing, China

&

Asif Mir

Department of Bioinformatics and Biotechnology,
International Islamic University, Islamabad, Pakistan

Quick Guideline for Computational Drug Design

Authors: Sheikh Arslan Sehgal, Rana Adnan Tahir, A. Hammad Mirza and Asif Mir

ISBN (eBook): 978-1-68108-603-3

ISBN (Print): 978-1-68108-604-0

© 2018, Bentham eBooks imprint.

Published by Bentham Science Publishers – Sharjah, UAE. All Rights Reserved.

First published in 2018.

General:

1. Any dispute or claim arising out of or in connection with this License Agreement or the Work (including non-contractual disputes or claims) will be governed by and construed in accordance with the laws of the U.A.E. as applied in the Emirate of Dubai. Each party agrees that the courts of the Emirate of Dubai shall have exclusive jurisdiction to settle any dispute or claim arising out of or in connection with this License Agreement or the Work (including non-contractual disputes or claims).

2. Your rights under this License Agreement will automatically terminate without notice and without the need for a court order if at any point you breach any terms of this License Agreement. In no event will any delay or failure by Bentham Science Publishers in enforcing your compliance with this License Agreement constitute a waiver of any of its rights.

3. You acknowledge that you have read this License Agreement, and agree to be bound by its terms and conditions. To the extent that any other terms and conditions presented on any website of Bentham Science Publishers conflict with, or are inconsistent with, the terms and conditions set out in this License Agreement, you acknowledge that the terms and conditions set out in this License Agreement shall prevail.

Bentham Science Publishers Ltd.
Executive Suite Y - 2
PO Box 7917, Saif Zone
Sharjah, U.A.E.
Email: subscriptions@benthamscience.org

BENTHAM SCIENCE

CONTENTS

FOREWORD

Bioinformatics is an interdisciplinary field which has played central role in the integration of computational techniques to solve biological problems. Computational methods have long been practiced in biology and perhaps, Fibonacci series was the most remarkable historical breakthrough in which Fibonacci used mathematical model for biological systems. Another exemplary work was performed by Gregor Mendel, who identified biological pattern of inheritance by devising Punnett square model for alleles of genes. Mathematical model integration was then continued by Thomas Hunt Morgan and later by his student who managed to group genes locus wise on chromosomes of Drosophila Melanogaster. Population geneticists used allelic frequency model to calculate frequency of most prevalent alleles in a population. With the sequencing of protein amino acids and DNA nucleotides, a new era of mathematical expansion in biology started. Margaret Dayhoff in 1978 designed Point Accepted Mutation (PAM) matrices to find out replacements of amino acids in primary structure of protein through the process of Natural Selection. Smith-Waterman and Needleman-Wunsch algorithms are key algorithms in alignment of gene sequences. In 1992, Henikoff and Henikoff presented Blossum matrix scoring function for protein alignment. Together, these alignment algorithms have currently become basics or starting point of most of the biological experiments. The alignment algorithms have helped us to discover not only mutational differences between sequences of genes but also these algorithms have been implemented to draw evolutionary relationships between living organisms. Martin Karplus, Michael Lavitt, and Arieh Warshel jointly shared 2013 Noble Prize for successful implementation of Mathematical simulation models to biological systems. They started a new era of Structural Biology that tends to find out 3-D structures of Biomolecules and their interactions with each other computationally. The accuracy of simulations is more than 90% that makes it a favorable choice of experiment when X-ray crystallography or NMR data is not available. Molecular docking is another technique that is employed to find out interactions between two biomolecules (preferably Protein-Protein Interactions or Protein-small molecule interactions) with minimum free energy state. Molecular docking has started new era of Computational Biology called Computer Aided Drug Design to design drugs against fatal diseases in exponentially less time.

"Far too many diseases do not have proven preventions or treatments. To make a difference for the millions of Americans who suffer from these diseases, we must gain better insights into the biological, environmental and behavioral factors that drive these diseases. Precision medicine is an emerging approach for disease treatment and prevention that takes into account individual variability in environment, lifestyle and genes for each person." (On January 20, 2015, President Obama announced the Precision Medicine Initiative® – PMI).

This era belongs to Personalized Medicine or Precision Medicine which collectively integrates all the Bioinformatics methods and analyses techniques including DNA sequences, family history, medical history, and environment to design individual specific medicines that will surely revolutionize lifestyles, lifespan, and thus Medical Science to its new heights.

As Bioinformatics methods and techniques are now considered as basic requirements for most of biological experiments, this book will help researchers to have an idea of how to initialize, design and add new layers of knowledge. This book does not cover all, but most of the major domains and utilizes most cited and most used tools and techniques. The material of book has been designed for novices to take them up to pro level in Bioinformatics. Many small tricks have also been discussed that would help researchers to easily understand theory, perform experiment, and analyze results. As a matter of fact, it is believed that this book must be part of every laboratory that needs Bioinformatics analyses for their experiments.

Muhammad Ismail, PhD
Director,
Institute of Biomedical and Genetic Engineering (IBGE)
Islamabad, Pakistan

PREFACE

The discussion of the bulk amount of informational data in biological sciences is still an understatement. The bioinformatics studies are being practiced in all over the world by universities, scientific groups, international and national companies and consortia, and it would not be underestimating if bioinformatics will be thought of as bedrock of future biological sciences. Bioinformatics evolved into complete interdisciplinary science to solve the biological problems by utilizing statistical, mathematical and computational approaches. Furthermore, the recording, annotation, storage, analyses, and searching/retrieval of biological information and representing this concluded information in order to understand various pathways, to educate ourselves, to understand biological processes in disease and healthy states, and also find potential drugs are only some aspects of bioinformatics. The conjuring of new 'Big-Data' associated typically with enormous databases of protein/gene sequences, and functional/structural information into which novel proteins and genes sequences are constantly deposited, can easily be searched by remote computer access with already known sequences. Numerous bioinformatics tools are freely available on the Internet to whoever wants to utilize them and it is impossible to list them all. The ability to collect, save, annotate, analyze, and distributes the biological information retrieved from sequencing and functional analyses is significant for modern biology. This is the reason behind the wide spread belief that the long run significance of bioinformatics approaches lies not much in the available tools, but in the annotation of data and concluding logical information that bioinformatics delivers for the improvement of life sciences.

Bioinformatics is to solve the biological problems with advanced computational manners by applying statistical and mathematical approaches. Structural bioinformatics is to predict and analyze 3D structures of macromolecules while Computer aided drug designing (CADD) helps scientists to design effective active molecules against diseases. However, the concept of structural bioinformatics including CADD is still hard to grasp for students and even more for educated laymen.

As a beginner, I fall into the first category with regards to this subject. Indeed, I first learn all these guidelines and tools for myself; in order to learn new emerging trends and approach my work with greater awareness about computational skills in biological problems. I felt that I need better insight into the CADD to achieve and improve the drug designing to reduce the time phase and resources. When I consulted the literature on this topic, I found that although there are many papers and books on the subject, the exact step by step working of basic tools and thorough analyses of generated results that I was looking for were scattered throughout them. I then began taking notes, and when I was finally satisfied, I discussed this with my colleague and a good friend (Rana Adnan Tahir). We realized after organizing the data in a suitable presentation, that it was potentially interesting for students and new learners to have visually detailed step by step working of tools and identify a lead compound. We believe that a better understanding of these concepts requires a more satisfactory verbal explanation than is visually provided, since, in our opinion, a visual step by step approach is the one closer to the understanding capability of non-experts and even for our own students. This is why this book is focused on providing a visual account and step by step guideline to identify a lead compound by computational means. In this sense, we deem that, besides the verbal working of tools, a visual step by step explanation can be very useful regarding designing of the compounds for the people who do not have a full understanding of it yet. Thus, we eventually came up with the present work, targeted to students and non-experts who are specifically interested in this domain and have a basic knowledge of structural bioinformatics.

With this book, we attempt to show the step by step working of basic tools for drug designing wherein analyses of generated results is always a priority. The protein sequence of RASSF2 was retrieved from UniProt and primary sequence analyses, secondary structure prediction and 3D structure prediction was performed followed by molecular docking studies. The basic information about bioinformatics and utilized tools is summarized in the first chapter. The book is summarized just as we wish we had been taught the core concepts and the working when we were first introduced to it. Therefore, we hope that it can be helpful as a guideline for drug designing by *in silico* approaches. The main text of the present work aims to demonstrate the step by step visual demonstration of the basic tools of structural bioinformatics and also is focused on the scientific point of views. If a beginner retrieves a protein sequence and performed all the steps described visually in this book will not only learn the computational drug designing but also can generate a publishable data.

Sheikh Arslan Sehgal
Department of Biosciences,
COMSATS Institute of Information Technology,
Sahiwal, Pakistan

State Key Laboratory of Membrane Biology,
Institute of Zoology; Chinese Academy of Sciences,
Beijing, China

University of Chinese Academy of Sciences,
Beijing, China

DEDICATION

"Thanks to ALMIGHTY ALLAH (GOD),
for the gift of writing he gave me".
S.A. Sehgal.

I know, you read many books in your life and you flip to the dedication page and find that the author has dedicated the book to someone else, not to you because the authors of those books were not your sons. But, in this book your son is the author and dedicating this book to you. I dedicate this book to my Father, Muhammad Ashraf Sheikh, who took me to the school for education, who always encouraged me to read books, and helped me transform into the person that I am.

To my Mother, Naeema Ashraf Sheikh, who always made delicious food for me.

To my Wife, Nisha Arslan Sehgal, whom I got married at the time of finalizing this book and spend more time on the laptop and she showed patience when her husband was working on weekends while on vacations too.

To my Brother, Babar Ashraf Sheikh, who made me an author, and always listened to my all stories from childhood uptil now.

To my Father-in-law, Chaudhary Muhammad Akram, who loves me a lot.

Sheikh Arslan Sehgal
Department of Biosciences,
COMSATS Institute of Information Technology,
Sahiwal, Pakistan

State Key Laboratory of Membrane Biology,
Institute of Zoology; Chinese Academy of Sciences,
Beijing, China

University of Chinese Academy of Sciences,
Beijing, China

ACKNOWLEDGEMENTS

We would like to acknowledge our students Zunaira Khalid, Sonia Kanwal, Naima Javed, Tassaduq Hussain Jaffar, Bilal Chaudhry, Nadeem Amin, Sumaiya Fatima and Azka Ahmed for their help in completing this project. The authors are also grateful to the Department of Biosciences, COMSATS Institute of Information Technology, Sahiwal, Pakistan for encouraging and providing the platform.

Sheikh Arslan Sehgal and Rana Adnan Tahir acknowledge their good friends Sajjad Ahmed Larra and Muhammad Sohail Raza for their kind support in preparing this book. Sheikh Arslan Sehgal acknowledges Dr. Ashir Masroor who becomes the ignition in this project.

CONFLICT OF INTEREST

The authors declare no conflict of interest, financial or otherwise.

Quick Guideline for Computational Drug Design

## CHAPTER 1

Introduction to Structural Bioinformatics

Abstract: Bioinformatics is an emerging modern biology to solve the biological problems with the help of computational approaches by utilizing mathematical and statistical techniques. There are many databases of protein and gene sequences. The protein sequence can be retrieved from UniProt and 3D structure prediction can be performed by utilizing homology modeling, threading and *ab initio*. The predicted structures can be evaluated by utilizing numerous model evaluation tools.

Keywords: Bioinformatics tools, Evolutionary biology, Homology modeling, Molecular docking, Sequence retrieval, Structural bioinformatics, Virtual screening.

MOTIVATION

The purpose of this chapter is to introduce readers to the basic views about bioinformatics and some basic concepts about bioinformatics in Computer Aided Drug Designing (CADD). The basic major approaches and the availability of tools and servers will also be explored in this chapter. The foremost step of most of the computational analyses is the acquiring of sequence and alignment of the protein of interest. This Chapter will also sketch the portrait of the complete book.

On 5th of October 1981, an article titled "Next Industrial Revolution: Designing Drugs by Computer at Merck" was published in the Fortune magazine (Van Drie, 2007). It is considered as the first step on CADD. Till now, many developments have been done in CADD and the High-throughput screening (HTS) technique also helped to develop the novel therapeutic targets. This technique depends on the automation for the screening of millions of molecules to find biologically active compounds. The advantage of this technique is to screen large libraries and scrutinize the biologically active molecules. The HTS technique usually results in numerous hits and some have the capability of becoming lead molecule despite the hits rate being low. From the last decade, number of biologically active compounds have been reported by applying the CADD. This approach reduces the workload and the time phase.

Here, we want to mention that the computational methods utilize the mathematical models for solving the biological problems. The tools and mathematical models discussed in this book are based on experimental data and have been tested for positive results. Due to an enormous increase in understanding of the biological world and experimental techniques, huge amounts of data is being produced daily. The mathematical models help to convert this raw data into useful information for the betterment of human health. Each tool has its own unique idea that is based on some natural rule found in available biological data. As mathematical models are based on statistical analysis, it is true that they don't give exact results, but for sure limits the search area for finding effective biological solutions. For the purpose of accuracy comparison, currently no standard has been set for measuring the accuracy of computational tools with traditional approaches. While the reported computational analyses showed reliable results and also proved by experimental work. *In silico* approaches and bioinformatics analyses have shown success in research methodologies to solve the biological problems (Sehgal *et al.*, 2013) and designed numerous novel computer-aided molecules against neurological disorders (Sehgal *et al.*, 2014, Sehgal *et al.*, 2015, Sehgal *et al.*, 2016, Sehgal 2017, Sehgal 2017, Khattak *et al.*, 2017) and cancer (Tahir *et al.*, 2013, Sehgal *et al.*, 2014, Kanwal *et al.*, 2016). *In silico* approaches and tools also applied to plants for logical conclusions (Sehgal *et al.*, 2012, Sehgal *et al.*, 2013, Zaka *et al.*, 2017).

BIOINFORMATICS

Bioinformatics is an interdisciplinary field that develops methods and tools for the understanding of biological data. As it is an interdisciplinary field of science, so the techniques of computer sciences, statistics, mathematics and engineering are used to process the biological data (Pan *et al.*, 2003). Bioinformatics is an emerging scientific field that utilizes computational, mathematical and statistical approaches for *in silico* solutions of biological problems (Sehgal *et al.*, 2013).

Here, in the beginning of this book, we want to explain that the bioinformatics is a vast field and have uncountable tools, software, servers, databases and also various approaches and techniques to perform the same task. We tried to describe the highly cited and effective approaches in this book and also try to elaborate in a simple manner for the understanding of the layman and the beginner.

Bioinformatics Approaches

Bioinformatics tools including software's for mapping, analyzing and aligning the DNA and protein sequences for analyses, creating and viewing the three dimensional (3D) models of protein structures. According to the basic approaches of bioinformatics, there are two fundamental ways of modeling biological systems *i.e.*, Static and Dynamic (Baldi *et al.*, 2001).

Static

It includes the sequences of proteins, nucleic acids and peptides. It also contains the interaction data for static entities including microarray data, networks of proteins and metabolites.

Dynamic

It includes the structures of nucleic acids, ligands (including metabolites and drugs) and peptides. The structures studied by using the bioinformatics tools are not considered static anymore. Systems Biology also comes under this category including reaction fluxes and variable concentrations of metabolites. The multi-agent-based modeling approaches capturing cellular events such as signaling, transcription and reaction dynamics are also considered as the dynamic approach of bioinformatics.

There are numerous sub-fields of bioinformatics but in this effort, the main focus will be on structural bioinformatics.

Structural Bioinformatics

Protein structure prediction is an important application of bioinformatics. It is the prediction of protein structure from its amino acid sequence. It helps to predict the folding, secondary, tertiary and quaternary structures of the target protein from its primary structure. Structure prediction is fundamentally different from the inverse problem of protein design. The amino acid sequence of a protein (primary structure) can be easily determined from the sequence of the gene that codes for it. Primary structure uniquely determines a structure in its native environment. The knowledge of primary structures is vital to understand the function of proteins. Structural information is usually classified as one of the secondary, tertiary and quaternary structures (Hogeweg *et al.*, 2011). A viable general solution to such predictions remains an open problem. Most of the efforts so far have been directed towards heuristics that help to solve these problems computationally.

In the genomic branch of bioinformatics, homology (study of distant relatives) is used to predict the function of a gene: if the sequence of gene A, whose function is known, is homologous to the sequence of gene B, whose function is unknown, one could infer that B may share A's function. In the structural branch of bioinformatics, homology is used to determine which part of a protein is important in structure formation and interaction with other proteins. In homology modeling approach, the structural and functional information is used to predict the protein structure by using the structural information of homologous proteins (Rodriguez *et al.*, 1998).

Software and Tools

Software, tools, databases and servers for bioinformatics range from simple command-line tools to more complex graphical programs and standalone web-services are also available by various bioinformatics companies and public institutions.

Open-Source Bioinformatics Software

Many free and open-source software and tools are available through public and private institutes while they continue to grow since the 1980s. Still, there is a need to develop new algorithms for the analyses of emerging biological problems. The open source tools often act as incubators of ideas, or community-supported plug-ins in commercial applications (Rodriguez *et al.*, 2010).

Web Services in Bioinformatics

The basic bioinformatics services are classified by the EBI into three categories: SSS (Sequence Search Services), MSA (Multiple Sequence Alignment) and BSA (Biological Sequence Analysis). The availability of these service-oriented bioinformatics resources demonstrates the applicability of web-based bioinformatics solutions and range from a collection of standalone tools with a common data format under a single, standalone or web-based interface, to integrative, distributed and extensible bioinformatics workflow management systems (Robert *et al.*, 2009).

Homology Modeling

Homology modeling is a method to construct an atomic-resolution model of the "target" protein from its amino acid sequence by utilizing an experimental 3D structure of a related homologous protein (the "template"). Homology modeling helps to identify one or more known protein structures having resemblance with the structure of the query sequence. The sequence alignment aligns the similar and identical residues of the query sequence with the template sequence. The protein structures are more conserved as compared to protein sequences amongst homologues but sequences falling below a 20% sequence identity can have very different structures (Chothia *et al.*, 2001). The sequence alignment and template structure are used to produce a 3D structural of the target protein (Marti *et al.*, 2000).

Steps of Homology Modeling

The homology modeling procedure can be broken down into four sequential steps: template selection, target-template alignment, model construction and model assessment (Marti *et al.*, 2000).

Template Selection

The critical step in homology modeling is the identification of the suitable template structure, if they are available. Template identification relies on the serial pair-wise sequence alignments provided by the database search techniques such as FASTA and BLAST.

Target-Template Alignment

Homology modeling is based on the alignment of target protein against the selected template protein. A template protein is an experimentally determined 3D structure obtained by X-ray crystallography or NMR techniques (Marti *et al.*, 2000). BLAST and ClustalW are the highly cited alignment tools for the alignment of protein sequences (Chenna *et al.*, 2003).

Model Construction

The aligned target sequence against the template protein sequence and the information obtained from alignment employed to generate the 3D structure of target protein, represented as a set of Cartesian coordinates for each atom in the protein (PDB format) (Baker *et al.*, 2001).

Model Assessment (Validation)

The validation of the predicted structure is another important step of homology modeling. The evaluation of the predicted structures validates the accuracy of the structures. On the basis of experimentally designed structures, numerous evaluation tools including PDBsum generate, rampage, ERRAT and verify3D are reported to validate the structures. The final selection of the predicted structures among all is based on the physiochemical properties of the protein, the cell type of protein expression and from the available data (Clarke *et al.*, 2000).

Accuracy

The accuracy of the structures depends on the sequence identity between the target and the templates sequences. At high sequence similarity, the primary source of error in homology modeling derives from the choice of the template or templates on which the model is based, while >40% sequence similarity exhibits serious errors in sequence alignment that inhibit the production of high-quality models (Venclovas *et al.*, 2005).

The sequence identity <30% can cause serious errors which results in the misprediction of the folds (Baker *et al.*, 2001). The similarity should be above 65% between the target and template for conclusive results.

Importance of Homology Modeling

The predicted structures are used in numerous bioinformatics analyses including protein protein interaction prediction, protein–protein docking, molecular docking, and functional annotation of genes identified in an organism's genome (Gopal *et al.*, 2001). Even low-accuracy homology models can be useful for these purposes because their inaccuracies tend to be located in the loops on the protein surface, which are normally more variable even between closely related proteins. The functional regions of the protein, especially its active site, tend to be more highly conserved and thus more accurately modeled (Baker *et al.*, 2001).

Homology models can also be used to identify subtle differences between related proteins that have not all been solved structurally. For example, the method was used to identify the cation binding sites on Na^+/K^+ ATPase and to propose hypotheses about different ATPases' binding affinity (Ogava *et al.*, 2002). By using in conjunction with molecular dynamics simulations, homology models can also generate the hypothesis about the kinetics and dynamics of a protein, as in studies of the ion selectivity of a potassium channel.

Homology Modeling Tools

Several tools are available for the homology modeling of proteins starting from structure prediction and go beyond molecular docking and molecular dynamic simulation. These tools can help to predict the structure of the respective protein as well as validate the model. The foremost step in homology modeling is sequence retrieval.

Sequence Retrieval

There are many ways and numerous databases are available to retrieve the sequence of the specific protein. UniProt Knowledgebase (UniProtKB) is one of the reliable database for sequence retrieval of proteins.

UniProtKB

The UniProtKB is the central hub for the collection of functional information on proteins, with accurate, consistent and rich annotation. In addition to capturing the core data mandatory for each UniProtKB entry (mainly, the amino acid sequence, protein name or description, taxonomic data and citation information), as much annotation information as possible is added. This includes widely accepted biological ontologies, classifications and cross-references and clear indications of the quality of annotation in the form of evidence attribution of experimental and computational data.

Here, we selected a protein RASSF2 (potential tumor suppressor gene acts as a KRAS-specific effectors protein) and will use the selected protein in all the analyses.

The search bar of UniProt was utilized (Fig. **1**) for moving towards the retrieval of the protein sequence. A page containing all the entries for RASSF2 appeared (Fig. **2**) and the respective species; *Homo sapiens* (Human) was selected.

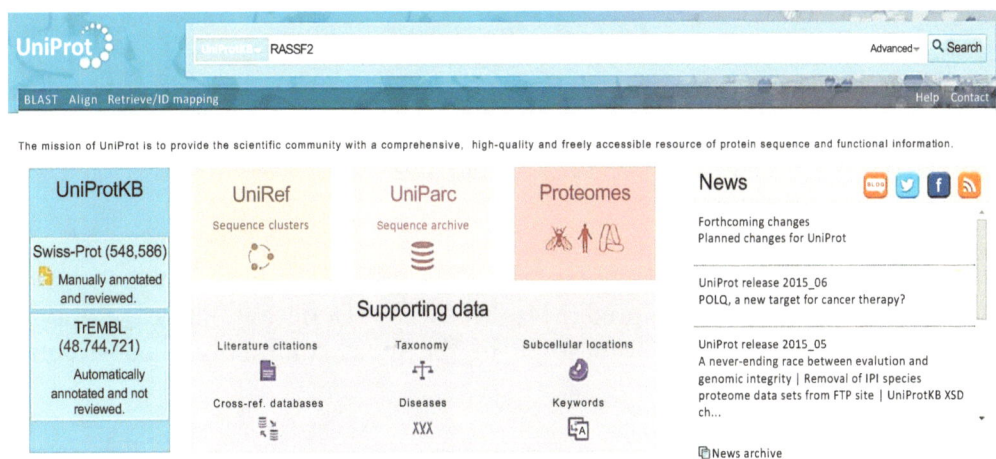

Fig. (1). Sequence retrieval for RASSF2 from UniProtKB.

It is better to get the 'Reviewed' sequences because they are highly curated according to the experimental data. In 'Reviewed' entries you need to know which organism's sequence is required for your experiment (Fig. **2**). The next step is to run protein blast for the RASSF2 fasta sequence obtained from UniProtKB having entry number P50749. For this purpose, either you can get fasta sequence of the selected protein, or you can just directly copy sequence of the selected protein, or you can only provide UniProt ID to blast program.

Characterization of Homology Modeling Tools

Homology Modeling tools are characterized as: (Here, we want to mention again that we tried to summarize and describe highly cited top-ranked freely available tools for all the study).

Fig. (2). UniProtKB showing different entries for RASSF2. 1) Search string will appear in search text box. 2) It is better to always choose the reviewed sequences for the further experiment because reviewed sequences are highly curated, validated and matched computationally with experimental and literature data. 3) You can either select the specific organisms from left bar, or you can choose one from the main screen.

Structure Prediction Tools

I-TASSER, Swiss-Model, HHpred, M4t, Modweb, 3D-Jigsaw, Phyre2, RaptorX, MODELLER and IntFOLD2 are some of the most commonly used structure prediction tools.

Model Validation Tools

ERRAT, Verify-3D, RAMPAGE, ANOLEA, Mol-probity the PDBsum Generate (Procheck) are widely used tools for model validation and evaluation (assessment).

Sequence Analysis Tools

We selected ProFunc and ProtParam Expasy for primary sequence analysis and secondary structure analysis respectively in this book.

Model Visualization Tools

UCSF Chimera, PyMol and VMD are highly intensive and extensible model visualization and analysis tools. These tools help to visualize the density maps, supra-molecular assemblies, sequence alignments, docking results, trajectories, and conformational ensembles. The high-quality images and animations can also be generated (Bartesaghi *et al.*, 2013). A quick look at Homology Modeling with Tools and software are mentioned in Fig. (**3**).

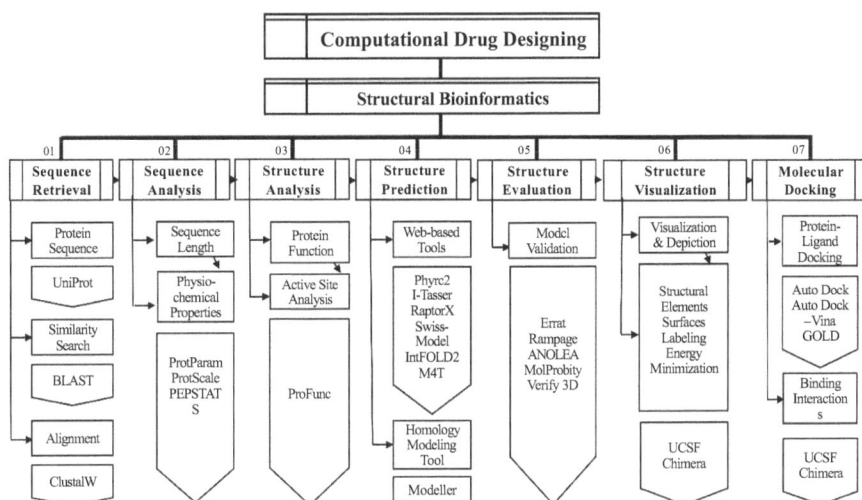

Fig. (3). A quick look at CADD with Tools and software.

Virtual Screening and Docking

Virtual Screening (VS) is a computational technique used in drug discovery to search libraries of small molecules to identify the potent molecules. VS can also evaluate large libraries of compounds by using computer programs. There are several databases and filters to filter the libraries for VS (PyrX, FILTER, LigandScout). Molecular docking is a method which predicts the preferred orientation of one molecule to a second when bound to each other to form a stable complex (Lengauer *et al.*, 1998).

REFERENCES

Alkhalifeh, K., Sarkis, R., Craeye, C. (2012). Design of a novel 3D circular Vivaldi antennas array for Ultra-wideband near-field radar imaging. *Proceedings of 6th European Conference on Antennas and Propagation, EuCAP 2012,* 898-901.
[http://dx.doi.org/10.1109/EuCAP.2012.6205930]

Altschul, S.F., Madden, T.L., Schäffer, A.A., Zhang, J., Zhang, Z., Miller, W., Lipman, D.J. (1997). Gapped BLAST and PSI-BLAST: a new generation of protein database search programs. *Nucleic Acids Res., 25*(17), 3389-3402.
[http://dx.doi.org/10.1093/nar/25.17.3389] [PMID: 9254694]

Apweiler, R., Martin, M.J., O'Donovan, C. (2012). Reorganizing the protein space at the Universal Protein Resource (UniProt). *Nucleic Acids Res., 40*(Database issue), D71-D75.

[http://dx.doi.org/10.1093/nar/gkr981] [PMID: 22102590]

https://doi.org/0039-6060(83)90109-5

Baker, C.C., Chaudry, I.H., Gaines, H.O., Baue, A.E. (1983). Evaluation of factors affecting mortality rate after sepsis in a murine cecal ligation and puncture model. *Surgery, 94*(2), 331-335.

Carvajal-Rodríguez, A. (2010). Simulation of genes and genomes forward in time. *Curr. Genomics, 11*(1), 58-61.
[http://dx.doi.org/10.2174/138920210790218007] [PMID: 20808525]

Chenna, R., Sugawara, H., Koike, T., Lopez, R., Gibson, T.J., Higgins, D.G., Thompson, J.D. (2003). Multiple sequence alignment with the Clustal series of programs. *Nucleic Acids Res., 31*(13), 3497-3500.
[http://dx.doi.org/10.1093/nar/gkg500] [PMID: 12824352]

Dayhoff, M.O. (1972). Atlas of Protein Sequence and Structure. *National Biomedical Research Foundation.*

Finkelstein, P.L., Ellestad, T.G., Clarke, J.F., Meyers, T.P., Schwede, D.B., Hebert, E.O., Neal, J.A. (2000). Ozone and sulfur dioxide dry deposition to forests: Observations and model evaluation. *J. Geophys. Res. D Atmospheres, 105*(D12), 15365-15377.
[http://dx.doi.org/10.1029/2000JD900185]

Gopal, S., Schroeder, M., Pieper, U., Sczyrba, A., Aytekin-Kurban, G., Bekiranov, S., Fajardo, J.E., Eswar, N., Sanchez, R., Sali, A., Gaasterland, T. (2001). Homology-based annotation yields 1,042 new candidate genes in the Drosophila melanogaster genome. *Nat. Genet., 27*(3), 337-340.
[http://dx.doi.org/10.1038/85922] [PMID: 11242120]

Gough, J., Karplus, K., Hughey, R., Chothia, C. (2001). Assignment of homology to genome sequences using a library of hidden Markov models that represent all proteins of known structure. *J. Mol. Biol., 313*(4), 903-919.
[http://dx.doi.org/10.1006/jmbi.2001.5080] [PMID: 11697912]

Harris, A.K., Bartesaghi, A., Milne, J.L., Subramaniam, S. (2013). HIV-1 envelope glycoprotein trimers display open quaternary conformation when bound to the gp41 membrane-proximal external-region-directed broadly neutralizing antibody Z13e1. *J. Virol., 87*(12), 7191-7196.
[http://dx.doi.org/10.1128/JVI.03284-12] [PMID: 23596305]

Hogeweg, P. (2011). The roots of bioinformatics in theoretical biology. *PLOS Comput. Biol., 7*(3), e1002021.
[http://dx.doi.org/10.1371/journal.pcbi.1002021] [PMID: 21483479]

Huelsenbeck, J.P., Ronquist, F. (2001). MRBAYES: Bayesian inference of phylogenetic trees. *Bioinformatics, 17*(8), 754-755.
[http://dx.doi.org/10.1093/bioinformatics/17.8.754] [PMID: 11524383]

Lengauer, C., Kinzler, K.W., Vogelstein, B. (1998). Genetic instabilities in human cancers. *Nature, 396*(6712), 643-649.
[http://dx.doi.org/10.1038/25292] [PMID: 9872311]

Martí-Renom, M.A., Stuart, A.C., Fiser, A., Sánchez, R., Melo, F., Sali, A. (2000). Comparative protein structure modeling of genes and genomes. *Annu. Rev. Biophys. Biomol. Struct., 29*(1), 291-325.
[http://dx.doi.org/10.1146/annurev.biophys.29.1.291] [PMID: 10940251]

Martí-Renom, M.A., Stuart, A.C., Fiser, A., Sánchez, R., Melo, F., Sali, A. (2000). Comparative protein structure modeling of genes and genomes. *Annu. Rev. Biophys. Biomol. Struct., 29*, 291-325.
[http://dx.doi.org/10.1146/annurev.biophys.29.1.291] [PMID: 10940251]

Nisbet, R., Elder, J., Miner, G. (2009). Handbook of Statistical Analysis and Data Mining Applications. Academic Press, Boston, Page 1-822, ISBN 9780123747655, https://doi.org/10.1016/B978-0-12-374-65-5.00045-0.

Ogawa, H., Toyoshima, C. (2002). Homology modeling of the cation binding sites of Na^+K^+-ATPase. *Proc Natl Acad Sci USA, 99*(25), 15977-15982.

Pan, X, Ning, W, Choi, A M (2003). Seeking of fuctional hypothetical genes from SAGE library. *Yangzhou Daxue Xuebao. Nongye Yu Shengming Kexue Ban, 25*(3), 29-32.

Rodriguez, R., Chinea, G., Lopez, N., Pons, T., Vriend, G. (1998). Homology modeling, model and software evaluation: three related resources. *Bioinformatics, 14*(6), 523-528. [http://dx.doi.org/10.1093/bioinformatics/14.6.523] [PMID: 9694991]

Sehgal, S. A., Tahir, R. A., Shafique, S., Hassan, M., Rashid, S. (2014). Molecular Modeling and Docking Analysis of CYP1A1 Associated with Head and Neck Cancer to Explore its Binding Regions. *Journal of Theoretical and Computational Science.*

Thompson, J.D., Higgins, D.G., Gibson, T.J. (1994). CLUSTAL W: improving the sensitivity of progressive multiple sequence alignment through sequence weighting, position-specific gap penalties and weight matrix choice. *Nucleic Acids Res., 22*(22), 4673-4680. [http://dx.doi.org/10.1093/nar/22.22.4673] [PMID: 7984417]

Venclovas, Č., Margelevičius, M. (2005). Comparative modeling in CASP6 using consensus approach to template selection, sequence-structure alignment, and structure assessment. [http://dx.doi.org/10.1002/prot.20725]

Sehgal, S., Khattak, N., Mir, A. (2013). Structural, phylogenetic and docking studies of D-amino acid oxidase activator (DAOA), a candidate schizophrenia gene. *Theor Biol Med Model, 10*(1), 3 DOI: 10.1186/1742-4682-10-3.

Sehgal, S.A. (2017). Pharmacoinformatics, Adaptive Evolution, and Elucidation of Six Novel Compounds for Schizophrenia Treatment by Targeting DAOA (G72) Isoforms. *Biomed Res Int, 2017*, 1–19 DOI: 10.1155/2017/5925714.

Sehgal, S.A. (2017). Pharmacoinformatics and molecular docking studies reveal potential novel Proline Dehydrogenase (PRODH) compounds for Schizophrenia inhibition. *Med Chem Res, 26*(2), 314–326 DOI: 10.1007/s00044-016-1752-2.

Sehgal, S.A., Mannan, S., Ali, S. (2016). Pharmacoinformatic and molecular docking studies reveal potential novel antidepressants against neurodegenerative disorders by targeting HSPB8. *Drug Des Devel Ther, 10*, 1605.

Sehgal, S.A., Mannan, S., Kanwal, S., Naveed, I., Mir, A. (2015). Adaptive evolution and elucidating the potential inhibitor against schizophrenia to target DAOA (G72) isoforms. *Drug Des. Devel. Ther, 9*, 3471.

Sehgal, S.A., Hassan, M., Rashid, S. (2014). Pharmacoinformatics elucidation of potential drug targets against migraine to target ion channel protein KCNK18. *Drug Des Devel Ther, 8*, 571.

Sehgal, S.A., Tahir, R.A., Shafique, S., Hassan, M., Rashid, S. (2014). Molecular Modeling and Docking Analysis of CYP1A1 Associated with Head and Neck Cancer to Explore its Binding Regions. *J Theor Comput Sci,*

Tahir, R.A., Sehgal, S.A., Khattak, N.A., Khattak, J.Z.K., Mir, A. (2013). Tumor necrosis factor receptor superfamily 10B (TNFRSF10B): an insight from structure modeling to virtual screening for designing drug against head and neck cancer. *Theor Biol Med Model. 10*(1), 1.

CHAPTER 2

Protein Primary Sequence Analysis

Abstract: The analyses of primary sequence (amino acid sequence) of protein always show logical insights for further studies. We can analyze a wide range of basic information including a prediction of repetitive sequences, the motif, domain, or active sites, and the ability to form a coiled-coiled structure by performing primary sequence analyses. Protparam, the primary sequence analysis tool was utilized to analyze the physicochemical properties of the selected protein.

Keywords: Amino acid sequence, Primary sequence, Protparam expassy.

MOTIVATION

The primary sequence analyses of a protein may be an effective start for the understanding of the protein of interest. There are certain features that are conserved in a protein family. The features including the length of the protein, amino acid compositions, hydrophobicity and ability to make α-helices or β-sheets provide useful information. The information gathered by visual analysis of primary sequence may include the prediction of repetitive sequences, prediction of the motif, domain, or active sites, and even its ability to form a coiled-coiled structure. There are many tools available for the primary sequence analysis including:

¬ ProtParam, for protein physic-chemical properties (available on Expasy).
¬ Compute pI/Mw, for calculating the pI and Molecular weight of the protein.
¬ ProtScale, for amino acid properties.
¬ RandSeq, for random protein sequence generation.
¬ PEPSTATS with Biochemistry-online provides Mass, pI, and percentage of the amino acids as acidic, basic or hydrophobic, and other related parameters at given pH.
¬ IPC calculates the isoelectric point of proteins at different scales.
¬ AllerTOP, AlgPred, SDAP, and SBS EpiToolKit can be used for the antigenicity properties of the protein and its related epitopes information.
¬ VIOLIN with VBLAST can be used in vaccine research experiments.

Sheikh Arslan Sehgal, Rana Adnan Tahir, A. Hammad Mirza & Asif Mir

Many other primary sequence analysis tools can be found at www.expasy.org and www.molbiol-tools.ca, here we only describe the functionality of ProParam tool from www.expasy.org.

In this Chapter, we will follow the methodology of section 02 mentioned in Fig. (**3**).

PROTPARAM EXPASY

ProtParam computes various physicochemical properties that can be deduced from a protein sequence. No additional information is required about the protein under consideration (Walker *et al.,* 2005).

Introduction

The parameters computed by ProtParam include the molecular weight, theoretical pI, amino acid composition, atomic composition, extinction coefficient, estimated half-life, instability index, aliphatic index and grand average of hydropathicity (GRAVY). The molecular weight and theoretical pI are calculated as in Compute pI/Mw. The amino acid and atomic compositions are self-explanatory (Gastegier).

Brief Instructions

Following instructions should be followed in order to evaluate the RASSF2 through ProtParam.

- Open web browser.
- Go to the ProtParam Expasy homepage (http://web.expasy.org/protparam/protpar-ref.html).
- Paste the FASTA sequence of the protein (RASSF2).
- Click on compute parameters.
- The new window will show the physiochemical properties of RASSF2.

Requirements

Input

The only input required for ProtParam is the FASTA sequence of the respective protein (RASSF2).

Sequence Submission

The FASTA sequence of RASSF2 is pasted in the box (Fig. **4**) or you can also use the accession number of sequence identifier of the protein.

Fig. (4). Protparam Expasy Sequence Submission Form.

Results Interpretation

The absorbance of light at a certain wavelength by a protein is the extinction coefficient of that protein. The aliphatic index of a protein is defined as the relative volume occupied by the aliphatic side chains. The GRAVY value shows the hydrophobicity of the protein, lower the GRAVY value will show the more hydrophobicity of the protein. The physicochemical properties of a RASSF2 protein are mentioned in Table **1**.

Table 1. Physicochemical properties of RASSF2.

Property	Value
No. of residues	326
Molecular Weight	37790.2
No. of Amino acid residues	326
Isoelectric Point (pI)	8.93
Extinction Co-efficient	28880-29005 M^{-1} cm^{-1}
Instability Index (II)	53.33
Aliphatic Index (AI)	82.76
GRAVY	-0.660
No. of negatively charged residues	42
No. of positively charged residues	47

Conclusion

ProtParam is useful for getting information about the primary sequence of the protein. The RASSF2 protein showed lower GRAVY value that depicts the hydrophobic nature of the protein. The protein showed >40 instability index predicts as unstable protein as RASSF2 showed 53.33.

Limitations of Primary Sequence Analyses

ProtParam Expassy

- It is not possible for ProtParam to specify the post translational modifications in query protein.
- It doesn't know whether the query mature proteins form multimers or dimers.
- It computes the results either by using the N-terminal amino acid or compositional data.
- It calculates the limited number of chemical and physical parameters of the query sequence that needs to be extend.
- Different results require Rasmol or Jmol for visualizations.

The major features, of ProtParam along with working, weblink and bibliography are mentioned in Table **2**.

Table 2. Analyses of primary sequence of protein

Tool Name	Availability	Function	Work Flow	URL	Reference
ProtParam Expasy	Web-based server	Computes various chemical and physical parameters of proteins. **Parameters:** Molecular weight, Theoretical pI, Amino Acid composition, aliphatic index, and grand average of hydropathicity (GRAVY) etc.	+Sequence Submission =Properties Prediction	http://web.expasy.org/protparam/	Gasteiger *et al.*, 2005

REFERENCES

Gasteiger, E., Hoogland, C., Gattiker, A., Duvaud, S., Wilkins, M.R., Appel, R.D., Bairoch, A. (2005). Protein Identification and Analysis Tools on the ExPASy Server. *In The Proteomics Protocols Handbook* [http://dx.doi.org/10.1385/1592598900]

Liu, Y., Xia, X., Xu, L., Wang, Y. (2013). Design of hybrid β-hairpin peptides with enhanced cell specificity and potent anti-inflammatory activity. *Biomaterials, 34*(1), 237-250. [http://dx.doi.org/10.1016/j.biomaterials.2012.09.032] [PMID: 23046754]

Sharma, N., Kushwaha, R., Sodhi, J.S., Bhalla, T.C. (2009). *In silico* analysis of amino acid sequences in relation to specificity and physiochemical properties of some microbial nitrilases. *J. Proteomics Bioinform., 2*(4), 185-192. [http://dx.doi.org/10.4172/jpb.1000076]

## CHAPTER 3

Secondary Structure Analyses

Abstract: The analyses of a secondary structure of protein always show logical insights for further studies. α-helices and β-sheets are considered as the major elements of secondary structure which are based on hydrogen bond donor and acceptor residues interactions. The secondary elements may provide information regarding the behavior of the proteins. ProFunc tool utilized to analyze the secondary structure of RASSf2.

Keywords: α-helices, β-sheets, Motifs, ProFunc, Profunc, Secondary structure.

MOTIVATION

The secondary structure analysis of the protein is very much important while working on structure and function of the proteins. Major secondary structure elements are α-helices and β-sheets which are based on hydrogen bond donor and acceptor residues interactions. The prediction of secondary elements may provide the information regarding the behavior of the protein including whether the protein localizes on membrane, or extracellular and intracellular. In some cases, secondary structure prediction helps to identify family-wise properties of the proteins involved in a certain function. Prosite and Scansite tools can predict the protein secondary structure elements including motif and conserved domains through only the protein sequence. Here, we will only discuss secondary structure prediction through protein coordinate file (PDB file) by utilizing the ProFunc tool.

In this Chapter, we will follow the methodology of section 03 mentioned in Fig. (**3**).

SECONDARY STRUCTURE OF PROTEINS

The secondary structure of a protein is a general 3D form of local segments of biopolymers such as proteins and nucleic acids (DNA/RNA). Secondary structure prediction involves a set of techniques in bioinformatics that aims to predict the local secondary structures of the proteins and RNA sequences based on the knowledge of their primary structure, amino acid and nucleotide sequence, respectively (Heringa *et al.,* 2000).

Sheikh Arslan Sehgal, Rana Adnan Tahir, A. Hammad Mirza & Asif Mir

ProFunc

ProFunc server is used to identify the likely biochemical functions of a protein from its 3D structure. It uses a series of methods, including fold matching, residue conservation, surface cleft analysis and functional 3D templates, to identify both the protein's likely active site and possible homologues in the Protein Data Bank (PDB) (Laskowski *et al.*, 2005).

Brief Instructions

Following instructions should be followed in order to analyze the secondary structure of the proteins.

- Open web browser
- Go to the ProFunc homepage (https://www.ebi.ac.uk/thornton-srv/databases/ProFunc/)
- The coordinates should be submitted in PDB format
- Simply, select it with the file dialog which is activated by clicking on the browse button (choose file)
- Click on upload button
- The result will be sent through email

Requirements

Input

The input to ProFunc is a single file containing the coordinates of respective protein structure (PDB file). The PDB file is uploaded at the ProFunc homepage.

Sequence Submission

The PDB file of RASSF2 is uploaded to ProFunc homepage (Fig. **5**).

Accessing of Data

After uploading the PDB file of the protein to the ProFunc submission form, results will be sent through E-mail.

LIMITATIONS OF SECONDARY STRUCTURE ANALYSES TOOLS

ProFunc

- ProFunc processing sometimes take many hours to analyze the 3D structures, that acquires attention for cumulative processing speed.
- Structure analyses are mainly conducted on known 3D structures.

- Various runs take much time to complete the analysis.
- Analyses are stored for 3 months but stored in a partition that is not backed up. Analyses can be lost at any time.

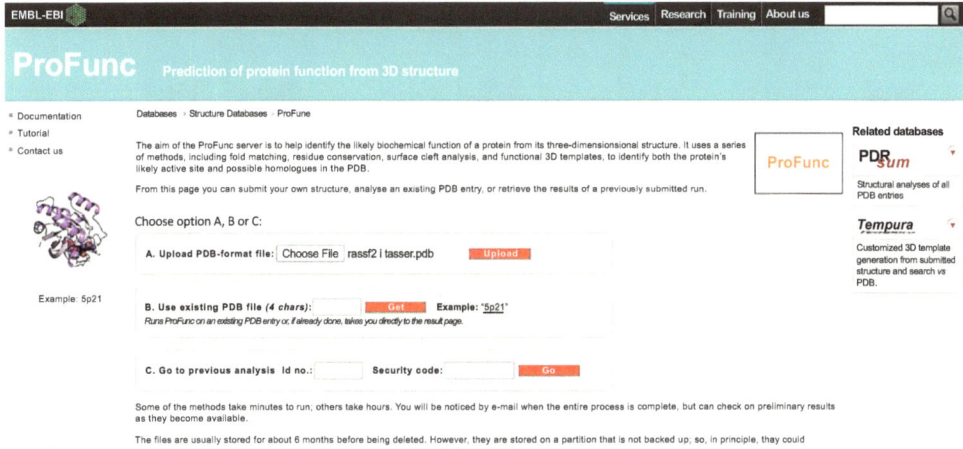

Fig. (5). ProFunc Submission Form.

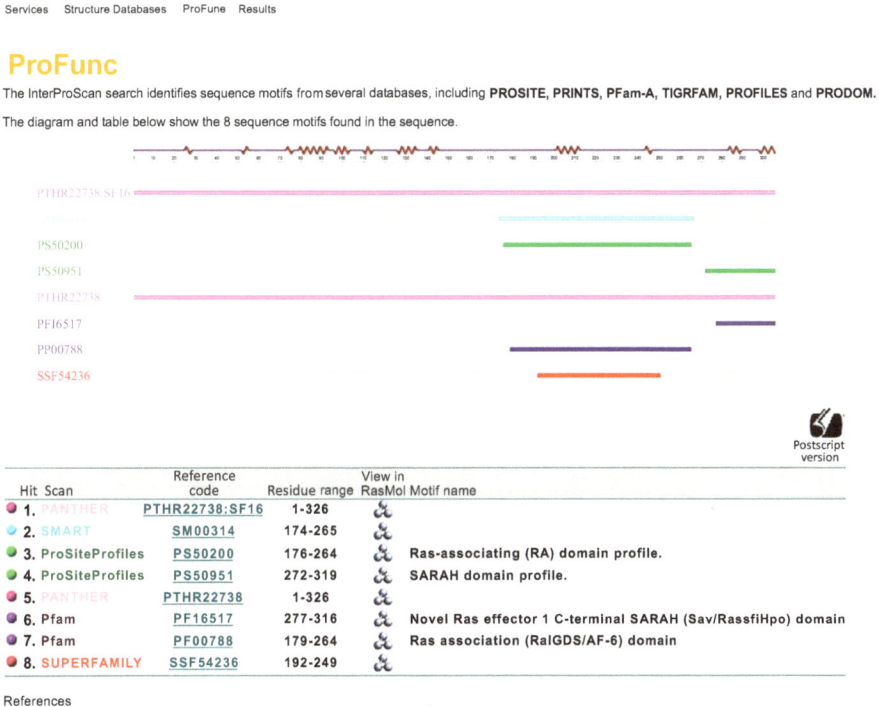

Fig. (6). ProFunc Results showing the Motifs identified through Interpro Scan Search.

ProFunc employs various methods comprising fold matching, alignments, cleft analysis, databases similarity searches to analyze the secondary structures summarized in Table **3**.

Fig. (7). ProFunc results showing alignments among query protein with proteins from PDB.

Table 3. Secondary Structure Analyses

Tool Name	Availability	Description	Work Flow	URL	Reference
ProFunc	Web-based Server	Structural analyses of 3d structures by various methods. **Methods:** fold matching, residue conservation, surface cleft analysis, and functional 3D templates etc.	+coordinates submission =Results through Email	https://www.ebi.ac.uk/thornton-srv/databases/ProFunc/	Laskowski *et al.,* 2005

REFERENCES

Brindha, S., Sailo, S., Chhakchhuak, L., Kalita, P., Gurusubramanian, G., Kumar, N.S. (2011). Protein 3D structure determination using homology modeling and structure analysis. *Sci. Vis., 11*(3), 125-133.

Laskowski, R.A., Watson, J.D., Thornton, J.M. (2005). ProFunc: a server for predicting protein function from 3D structure. *Nucleic Acids Res., 33*(Web Server issue) (Suppl. 2), W89-93.
[http://dx.doi.org/10.1093/nar/gki414] [PMID: 15980588]

Notredame, C., Higgins, D.G., Heringa, J. (2000). T-Coffee: A novel method for fast and accurate multiple sequence alignment. *J. Mol. Biol., 302*(1), 205-217.
[http://dx.doi.org/10.1006/jmbi.2000.4042] [PMID: 10964570]

## CHAPTER 4

3D Structure Prediction

Abstract: *The proteins having similar sequences may have similar structure*" is the philosophy of sequences prediction techniques. The 3D structures have structural and functional importance for drug designing. Homology modeling, threading and *ab initio* approaches are the computational techniques to predict the 3D structure of the proteins for further studies. In this chapter, we will precisely elaborate numerous tools of 3D structure prediction.

Keywords: *ab initio*, Homology modeling, I-Tasser, MODELLER, Structure prediction, Threading.

MOTIVATION

3D structure of a protein is important for the sequences and functional analysis of a protein. Experimentally determined structures through X-ray crystallography and Nuclear Magnetic Resonance (NMR) techniques provide detailed information related to the dynamics of the protein interactions with ligands, receptors, interacting proteins, and with its environment. There are 35,914 structures of human sequences available in PDB but still, usually the protein of interest may not have its structure solved through experimental techniques. In such conditions, it becomes necessary to have opted different methodologies for structural analyses. Computational methods are providing alternative ways to solve this problem by utilizing the structure prediction techniques. This chapter is aimed to introduce different desktop and web-based structure prediction methods. Many tools will be discussed and each tool implementing a different algorithm to model the protein structures. It will help the researchers and also the beginners to understand 3D structure prediction and can also compare the results according to the behavior of the protein of interest.

"The proteins having similar sequences may have similar structures" is the philosophy of structure prediction techniques. Due to unavailability of resources and time trade-offs, it is not possible to find the structure of proteins through X-ray crystallography and NMR techniques.

Sheikh Arslan Sehgal, Rana Adnan Tahir, A. Hammad Mirza & Asif Mir

Structure prediction tools help to solve the 3D structure of proteins based on similarity (homology) of template protein structure with query protein (protein of interest). All of the structure prediction algorithms try to first align sequences of the protein of interest with the template(s), so first most important step in structure prediction is alignment. In this Chapter, we will follow the methodology of section 04 mentioned in Fig. (**3**).

STRUCTURE PREDICTION TOOLS

Protein structure prediction is an effort to predict the 3D structure of a protein from its amino acid sequence. The primary sequence also used to predict the folding and its secondary, tertiary, and quaternary structure (Mount *et al.*, 2001).

A large number of protein structure predictions are very expensive and time consuming by X-Ray crystallography and NMR methods (Sehgal *et al.*, 2014). Numerous computational approaches and tools for 3D structure prediction have been developed. It is critical that the biological community and beginners are aware of such tools and is able to interpret their results in an informed way. The elaborative methodology may provide a guideline to predict the 3D structure by different approaches and also interpret the results effectively.

PHYRE2

Phyre2 is a structure prediction non-commercial tool and is able to regularly generate the reliable protein models (Lawrence *et al.*, 2011).

Introduction

Phyre2 server predicts the 3D structure of target protein sequence by using homology modeling approach. It has been ranked amongst one of the best systems of its kind for the last four years as judged by the biannual Critical Assessment of Structure Prediction (CASP) meetings (Lawrence *et al.*, 2011).

Brief Instructions

Following instructions should be followed in order to predict the 3D structure of the respective protein.

- Open the web browser
- Go to Phyre2 homepage (http://www.sbg.bio.ic.ac.uk/phyre2/html/page.cgi?id=index)
- An interface will open where the amino acid sequence of the respective protein is uploaded *e.g.* RASSF2
- Provide the institutional e-mail ID (abc@uni.edu.pk)

• Results will be sent through e-mail depending upon the sequence (query) length and number of protein sequences in queues

Requirements

Input

The input required for Phyre2 is the FASTA sequence of target protein retrieved from UniProt.

Sequence Submission

The FASTA sequence of RASSF2 submitted to phrex2 (Fig. **8**).

Fig. (8). RASSF2 Sequence submission to Phyre2.

Click on the "**Phyre search**" option and the generated 3D structure will send to the mentioned mail ID.

Results Interpretation (Results Screen)

The detailed information about the target protein (RASSF2) can be retrieved by opening the Phyre2 link sent to the user by e-mail containing results. The results screen of Phyre2 is divided into following main sections:

Secondary Structure and Disorder Prediction

Initially, the target protein sequence scanned against a large sequence database by employing PSI-BLAST. The predicted α-helices, β-strands and disordered regions are shown graphically together with a color-coded confidence bar (Fig. **9**).

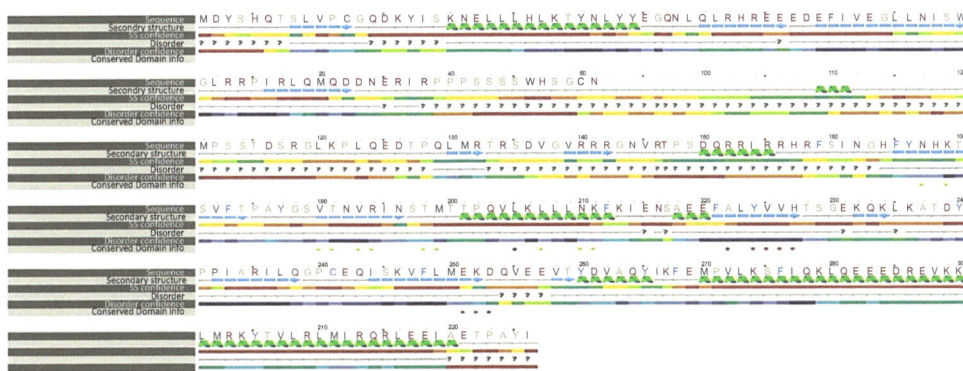

Fig. (9). Phyre2 output for secondary structure and disorder prediction.

Domain Analysis

Many proteins contain multiple domains. Phyre2 provides a table of template matches color-coded by confidence and indicating the region of the user sequence matched (Fig. **10**).

Fig. (10). Phyre2 output showing multiple domains and model viewer.

Detailed Template Information

The main results table in Phyre2 provides confidence estimates, images and links to the 3D predicted models and information derived from either Structural Classification of Proteins database (SCOP) or PDB depending on the source of the detected template. For each match, a link takes the user to a detailed view of the alignment between the target sequence and the template sequence 3D structure (Fig. **11**).

Fig. (11). Detailed template information table.

Alignment View

The detailed alignment view permits to examine the individually aligned residues, matches between target and template secondary structure elements and the ability to toggle information regarding the patterns of sequence conservation and secondary structure confidence. In addition, J-mol is used to visualize the 3D view of protein model (Fig. **12**).

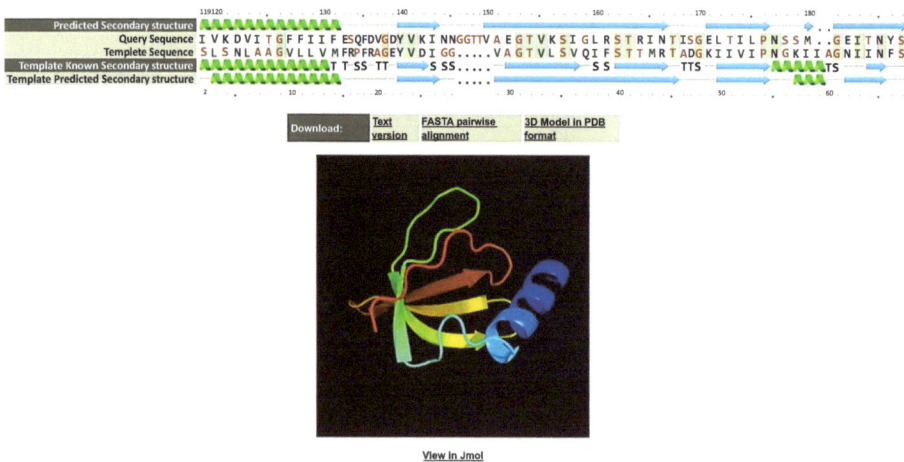

Fig. (12). Phyre2 detailed view of the alignment between a user sequence and known protein structure.

Conclusion

Phyre2 can be helpful for predicting the 3D structure, function prediction, domain prediction, domain boundary prediction, evolutionary classification of proteins, guiding site-directed mutagenesis and solving protein crystal structures by molecular replacements. The 3D structure of RASSF2 was also predicted by Phyre2.

I-TASSER

I-TASSER (Iterative Threading ASSEmbly Refinement) (Zhang *et al.*, 2010) is one of the top-ranked homology modeling/threading approach tool for 3D structure prediction.

Introduction

I-TASSER is an online tool for the 3D structure prediction of protein molecules from amino acid sequences. It detects templates from PDB by a technique called fold recognition (Zhang *et al.*, 2010).

Brief Instructions

Following instructions should be followed in order to predict the 3D structure of the target protein.

- Open the web browser
- Go to the I-TASSER homepage (http://zhanglab.ccmb.med.umich.edu/I-TASSER)
- Upload the amino acid sequence (RASSF2) or paste the sequence to the interface of I-TASSER
- Provide the institutional e-mail ID (abc@uni.edu.pk)
- Type the name of your current job (optional)
- Results will be sent through e-mail depending upon the sequence (query) length

Requirements

Input

The only input required by I-TASSER is the amino acid sequence of protein (RASSF2) in FASTA format.

Sequence Submission

The FASTA format sequence of the target protein which is RASSF2 here submitted to I-TASSER interface (Fig. **13**).

Fig. (13). Sequence Submission to I-TASSER.

Accessing Data

The submitted protein sequence will be processed by I-TASSER and the user have to wait for the results depends upon the sequence length, a number of homologous sequences, frequency and length of insertions and deletions for a prediction to complete. The summary information and the predicted structures in PDB format are sent to the user together with a link to a web page containing results through the mail.

Results Interpretation (Output)

Results of I-TASSER are divided into two sections.

Structure Prediction

Structure prediction helps to find the following information about RASSF2.

Top 10 Threading Templates Used by I-TASSER

I-TASSER uses the templates of the highest significance in the threading alignments, the significance of which is measured by the Z-score (the difference between the raw and average scores in the unit of standard deviation). Usually, one template of the highest Z-score is selected from each threading program, where the threading programs are sorted by the average performance in the large-scale benchmark test experiment (Zhang *et al.*, 2010). I-TASSER showed top ten templates for RASSF2 (Fig. **14**).

Rank	PDB Hit	Iden1	Iden2	Cov	Norm. Z-score	Download Align.
1	3ddcB	0.26	0.11	0.32	1.29	Download
2	5bq9A	0.08	0.20	0.94	1.38	Download
3	3ec8A	0.17	0.09	0.40	1.12	Download
4	4qiwC	0.09	0.18	0.90	1.03	Download
5	3ec8A	0.20	0.09	0.36	1.10	Download
6	4txdA	0.07	0.17	0.93	1.26	Download
7	3ec8A	0.18	0.09	0.24	2.15	Download
8	3txaA	0.08	0.17	0.91	1.26	Download
9	2m4nA	0.12	0.06	0.29	1.71	Download
10	3ddcB	0.25	0.11	0.40	0.74	Download

Fig. (14). Top ten threading templates used by I-TASSER.

Top 5 Final Model Predicted by I-TASSER

The confidence of each model is quantitatively measured by C-score and the calculation of C-score based on the significance of threading template alignments and the convergence parameters of the structure assembly simulations. C-score is typically in the range of [-5, 2], where a C-score of a lower value signifies a model with a lower confidence and *vice-versa*. RMSD and TM-score are estimated based on C-score and the protein length following the correlation observed between these qualities. The generated top 5 predicted structures were ranked by the cluster size. Although the first model usually consider as a better quality model, the possibility also stands that the lower-rank models have a better quality than the higher-rank models depends on the nature of the protein and the experiments. I-TASSER predicted and showed the top 5 models for RASSF2 (Fig. **15**).

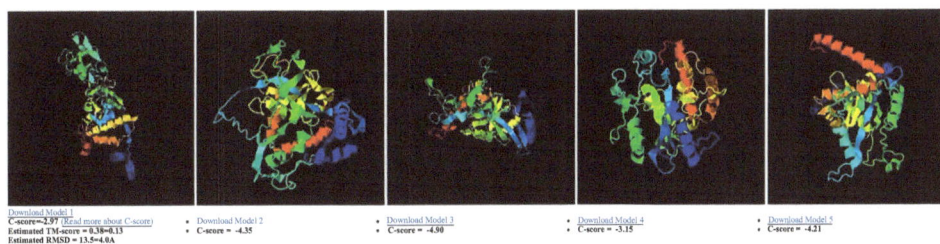

Download Model 1
C-score=-2.97 (Read more about C-score)
Estimated TM-score = 0.38±0.13
Estimated RMSD = 13.5±4.0Å

• Download Model 2
• C-score = -4.35

• Download Model 3
• C-score = -4.90

• Download Model 4
• C-score = -3.15

• Download Model 5
• C-score = -4.21

Fig. (15). Top 5 models for RASSF2 predicted by I-TASSER.

Structurally Closely Related Protein

The structure assembly simulation was also done to the predicted models and I-TASSER used the TM-align structural alignment program to match the first I-TASSER model to all the available structures in the PDB library. This matching program showed top 10 proteins hunted from the PDB that have the closest structural similarity. Due to the philosophy of structural similarity, the hunted proteins by matching program often showed similar function to the target protein. However, it is much better to use the data generated by the 'Predicted function using COACH' to infer the function of the target protein. COACH has been utilized to derive the biological functions from multi-source of the sequences and structural features and also showed reliable accuracy than the function annotations derived only from the global structure comparison (Fig. **16**).

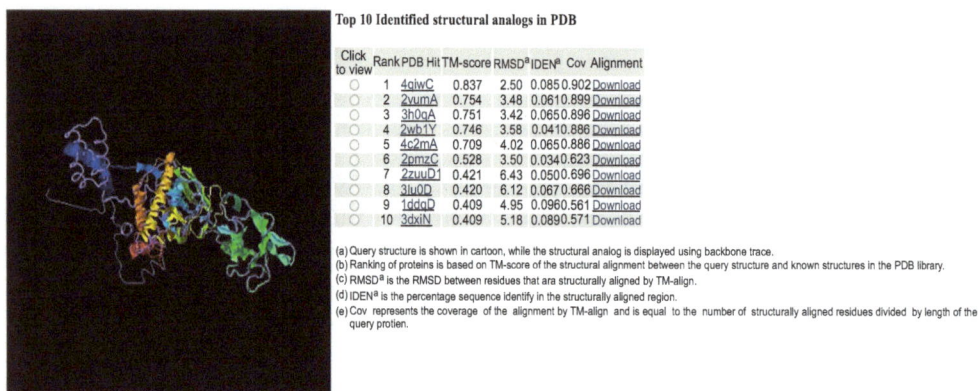

Top 10 Identified structural analogs in PDB

Click to view	Rank	PDB Hit	TM-score	RMSD[a]	IDEN[a]	Cov	Alignment
○	1	4giwC	0.837	2.50	0.085	0.902	Download
○	2	2vumA	0.754	3.48	0.061	0.899	Download
○	3	3h0qA	0.751	3.42	0.065	0.896	Download
○	4	2wb1Y	0.746	3.58	0.041	0.886	Download
○	5	4c2mA	0.709	4.02	0.065	0.886	Download
○	6	2pmzC	0.528	3.50	0.034	0.623	Download
○	7	2zuuD1	0.421	6.43	0.050	0.696	Download
○	8	3lu0D	0.420	6.12	0.067	0.666	Download
○	9	1ddqD	0.409	4.95	0.096	0.561	Download
○	10	3dxiN	0.409	5.18	0.089	0.571	Download

(a) Query structure is shown in cartoon, while the structural analog is displayed using backbone trace.
(b) Ranking of proteins is based on TM-score of the structural alignment between the query structure and known structures in the PDB library.
(c) RMSD[a] is the RMSD between residues that are structurally aligned by TM-align.
(d) IDEN[a] is the percentage sequence identify in the structurally aligned region.
(e) Cov represents the coverage of the alignment by TM-align and is equal to the number of structurally aligned residues divided by length of the query protien.

Fig. (16). Structural analogs for RASSF2.

Protein Function

It includes:

Ligand Binding Site

I-TASSER also predicts the ligand binding sites through biological annotations of the target protein by utilizing COACH. COACH is a meta-server approach that combines the multiple function annotation results from the S-SITE, TM-SITE and COFACTOR programs (Fig. **17**).

Conclusion

I-TASSER provides 5 models for RASSF2 that are helpful for evaluating the best model for further analysis. The predicted structures from I-TASSER has the high level of accuracy and is a highly cited software.

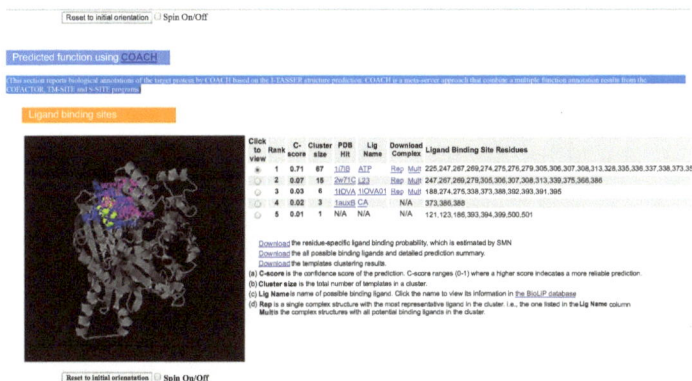

Fig. (17). I-TASSER results for RASSF2.

RaptorX

Xu group developed a tool named RaptorX to predict the 3D structures for the protein sequences without close homologs in PDB.

Introduction

RaptorX predicts the secondary and tertiary structures including binding sites, disordered regions, solvent accessibility and contacts of the target protein by utilizing the amino acid sequence. RaptorX also shows the confidence scores to indicate the quality of a predicted 3D model, P-value for the relative global quality of the predicted structure, global distance test (GDT) and un-normalized GDT (uGDT) for the absolute global quality, and modeling error at each residue. RaptorX also excels at the alignment of hard targets, which have less than 30% sequence identity with the solved structures in PDB (Xu *et al.,* 2011).

Brief Instructions

Following instructions should be followed in order to predict the 3D structure of the target protein.

- Open the web browser
- Go to RaptorX homepage
- Upload the FASTA sequence of the protein (RASSF2) or paste the sequence to the RaptorX submission form (http://raptorx.uchicago.edu/StructurePrediction/predict/)
- Provide the institutional e-mail ID (abc@uni.edu.pk)
- Type the name of your current job (optional) and submit the job
- Results will be sent by e-mail

Requirements

Input

The only input required for RaptorX is the FASTA sequence of the target protein (RASSF2).

Sequence Submission

The FASTA sequence of RASSF2 is pasted in the dialogue box provided by RaptorX interface (Fig. **18**).

Fig. (18). Sequence submission to RaptorX.

Submit the sequence by using "submit" option and the 3D models of RASSF2 will be received through the mail.

Accessing Data

The user will have to wait for the models depend on the number of homologous sequences, the frequency and length of insertions and deletions for a prediction to complete. The summary information and the predicted structure in PDB format will send to the user through the mail.

Results Interpretation (Output)

The predicted structures by RaptorX provided following results (sections) for RASSF2.

Input Sequence and Domain Partition

The domains of the protein sequence (RASSF2) were also predicted by RaptorX (Fig. **19**).

```
1           11          21          31          41          51          61          71          81          91
MDYSHQTSLV  PCGQDKYISK  NELLLHLKTY  NLYYEGQNLQ  LRHREEEDEF  IVEGLLNISW  GLRRPIRLQM  QDDNERIRPP  PSSSSWHSGC  NLGSQGTTLK
2222222222  2222222222  2222222222  2222222222  2222222222  2222222222  2222222222  2222222000  0000000000  0000000000
101         111         121         131         141         151         161         171         181         191
PLTVPKVQIS  EVDAPPWGDQ  MPSSTDSRGL  KPLQEDTPQL  MRTESDVGVR  RRGNVRTPSD  QRRIRRHRFS  INGHFYNHKT  SVFTPAYGSV  INVRINSTMT
0000000000  0000000000  0000000000  0000000000  0000000000  0000000000  0000000000  0000000111  1111111111  1111111111
201         211         221         231         241         251         261         271         281         291
TPQVLKLLLN  KFKIENSAEE  FALYVVHTSG  EKQKLKATDY  PLIARILQGP  CEQISKVFLM  EKDQVEEVTY  DVAQYIKFEM  PVLKSFIQKL  QEEEDREVKK
1111111111  1111111111  1111111111  1111111111  1111111111  1111111111  1111111333  3333333333  3333333333  3333333333
301         311         321
```

Fig. (19). Input sequence and domain partition.

Prediction Results

The predicted structure for RASSF2 provided by RaptorX can be downloaded from the result screen (Fig. **20**).

Detailed Prediction Results

The detailed results predicted for the target protein can be obtained from the output sent through the mail (Fig. **21**).

SWISS-MODEL

SWISS-MODEL is an automated web server to construct the 3D structures of the proteins, accessible *via* Expasy web server.

Introduction

SWISS-MODEL provides a personalized web-based area for each user in which the protein models can be built and the results are stored and analyzed (Marti-Renom *et al.,* 2006).

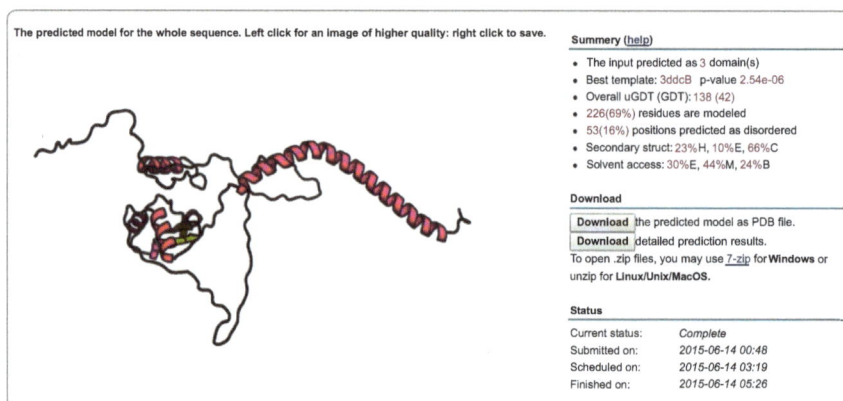

The predicted model for the whole sequence. Left click for an image of higher quality: right click to save.

Summery (help)

- The input predicted as 3 domain(s)
- Best template: 3ddcB p-value 2.54e-06
- Overall uGDT (GDT): 138 (42)
- 226(69%) residues are modeled
- 53(16%) positions predicted as disordered
- Secondary struct: 23%H, 10%E, 66%C
- Solvent access: 30%E, 44%M, 24%B

Download

Download the predicted model as PDB file.
Download detailed prediction results.
To open .zip files, you may use 7-zip for **Windows** or unzip for **Linux/Unix/MacOS.**

Status

Current status:	*Complete*
Submitted on:	*2015-06-14 00:48*
Scheduled on:	*2015-06-14 03:19*
Finished on:	*2015-06-14 05:26*

Fig. (20). Prediction 3D model of RASSF2 by RaptorX.

Section III. Detailed Prediction Results (you can see each result entry by clicking on it)

```
[+] Click to view Structure labeling for the whole sequence
[+] Click to view 3D model(s) for domain 1 [178, 267] P-value:2.54e-06
[+] Click to view 3D model(s) for domain 2 [1, 77] P-value:1.10e-02
[+] Click to view 3D model(s) for domain 3 [268, 326] P-value:3.90e-02
```

Fig. (21). Prediction Results.

Brief Instructions

Following instructions should be followed in order to predict the 3D structure of the target protein by using SWISS-MODEL.

- Open the web browser
- Go to SWISS-MODEL homepage (http://swissmodel.expasy.org/interactive)
- Upload the amino acid sequence of the target protein (RASSF2)
- Provide your mail ID
- Results will be sent through e-mail

Requirements

Input

The only input required for SWISS-MODEL is the FASTA sequence of the protein (RASSF2).

Sequence Submission

At SWISS-MODEL homepage, the FASTA sequence of the target protein

(RASSF2) submitted for the structure prediction (Fig. **22**).

Fig. (22). Sequence submission form.

Accessing Data

The amino acid sequence submitted to the SWISS-MODEL and the suitable templates list appeared and the models can be downloaded by clicking on the models.

Results Interpretation (Results Screen)

The results screen of SWISS-MODEL is divided into three main sections.

Summary

The summary showed the top 5 templates of the RASSF2 protein (Fig. **23**).

Templates

This section showed all the possible templates of the RASSF2 protein (Fig. **24**).

Models

The model's section showed the predicted models of the target protein and can be downloaded for further analyses (Fig. **25**) which we will discuss in later chapters.

Fig. (23). Top five templates.

Fig. (24). Template list.

Conclusion

Swiss-Model has provided the 3D predicted models for RASSF2. The models will be further verified and evaluated on different parameters for further studies.

3D-JIGSAW

This server also builds the 3D structures of the proteins based on homologues of known structure.

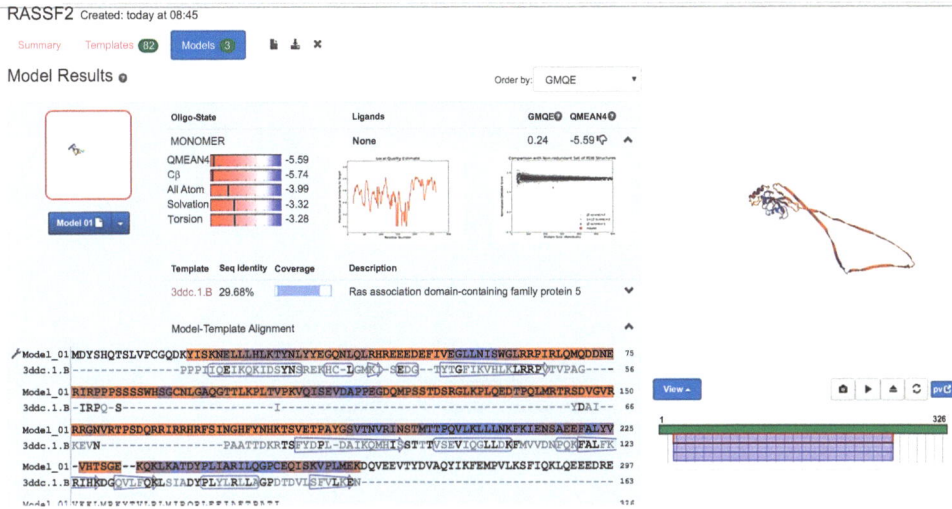

Fig. (25). The result screen containing 3D predicted model.

Introduction

3D-JIGSAW is an automated system to build the 3D structures and can either be run in completely automatic mode by means of a web server or individual modules of the program may be executed separately. The intermediate files saved which can be modified as per the need and nature of the protein and experiment (Tramontano *et al.* 2001).

Brief Instructions

Following instructions should be followed in order to predict the 3D structure of the target protein (RASSF2) by 3D-JIGSAW.

- Open the web browser
- Go to 3D-JIGSAW homepage
  (http://bmm.cancerresearchuk.org/ ~3djigsaw/html/form_v2.0huygens.html)
- Upload the amino acid sequence of the protein
- Provide your e-mail ID
- Submit the query
- Results will be sent through e-mail

Requirements

Input

The only input required for 3D-JIGSAW is the FASTA sequence of the protein

(RASSF2).

Sequence Submission

The amino acid sequence of the target protein submitted to 3D-JIGSAW as input
(Fig. **26**).

Fig. (26). Sequence submission form for 3D-JIGSAW.

After submitting the sequence to the tool, a new window will open having the
notification about the retrieval of a 3D predicted model for RASSF2 through the
mail (Fig. **27**).

Fig. (27). Notification for the submission of RASSF2 sequence to 3D-JIGSAW.

Accessing Data

The predicted structures by 3D-Jigsaw will be sent through given e-mail.

Results Interpretation

The PDB files of the predicted structures can be downloaded by the mail (Fig. 28).

lif-3djigsaw@crick.ac.uk   *Jun 14 at 1:52 PM*
To zunera.khalid93@yahoo.com

The 3D-JIGSAW model for your sequence is attached to this email, you can open it with a program like Rasmol or Pymol

        Thanks for using 3D-JIGSAW
The Francis Crick Institute Limited is a registered charity in England and Wales no. 1140062 and a company registered in England and Wales no. 06885462, with its registered office at 215 Euston Road, London NW1 2BE.

RASSF2.pdb   Download ⌄

Fig. (28). Getting 3D model for RASSF2 from email.

Conclusion

3D-Jigsaw predicted the 3D structures for RASSF2.

INTFOLD2

IntFOLD2 is integrated protein structure and function prediction server used to predict the 3D structures.

Introduction

IntFOLD2 predicts the 3D structure, assess the quality of predicted structure, ligand binding site residues, detect disordered regions and also predicts the boundaries for structural domains of the target amino acid sequence (McGuffin *et al.*, 2012).

Brief Instructions

Following instructions should be followed in order to predict the 3D structure of the target (RASSF2) protein.

- Open the web browser
- Open IntFOLD2 homepage (http://www.reading.ac.uk/bioinf/IntFOLD/ IntFOLD2_form.html)
- The amino acid sequence of the target protein submission
- Provide the institutional mail ID (abc@uni.edu.pk)
- Results will be sent through the mail depending upon the sequence (query) length and number of queries sent to IntFOLD2

Requirements

Input

The only input required for IntFOLD2 is the FASTA sequence of the target protein (RASSF2).

Sequence Submission

The amino acid sequence submitted to IntFOLD2 (Fig. **29**) for the structure prediction.

The IntFOLD Integrated Protein Structure and Function Prediction Server (Version 2.0)

This simple form allows you to: predict tertiary structures, assess the quality of 3D models, detect disordered regions, predict the bounderies for structural domains and predict likely ligand binding site residues for a submitted amino acid sequence.

Further information, news and references will be posted on the IntFOLD home page. Please refer to the help page before submitting any data. Click 'Help' in each section for detailed instructions.

Required - Input sequence of protein target *(single letter code)* Help

Optional - Upload model/models *(either a single PDB file or a tarred and gzipped directory of PDB files)* Help
Choose File | target rassf2.txt
Optional - Short name for protein target Help
RASSF2
Optional - E-mail address Help
abc123@ciit.edu.pk

● By selecting this button, I assert that I am part of an academic institution (not a government research lab such as the NIH, or a commercial entity) and I agree to the terms and conditions of the Modeller license.
○ I am a non-academic and I have a MODELLER access key:
Help, I need a Key!

Fig. (29). Sequence Submission to IntFOLD2.

Accessing Data

The user has to wait for the results that will be sent through the mail.

Results Interpretation (Output)

The PDB file can be downloaded from the e-mail.

Conclusion

IntFOLD2 predicted the 3D structure of RASSF2.

HHPRED

HHpred predicted the 3D structures of the target protein and detects the sensitive homologous proteins as templates and predicts the structure by using HMM-HMM-comparison.

Introduction

Initially, HHpred starts from a query sequence and generate a MSA by using HHblits and turns it into a profile HMM. The generated HMM profile of the target sequence compared with a database of HMMs representing the proteins with known structures (PDB, SCOP) or annotated the protein families (PFAM, SMART, CDD, COGs, KOGs). The output templates are the closest homologs to alignments. HHpred can also predict the 3D homology models by using the identified the templates from PDB. It can optimize the template picking and query-template alignments for homology modeling (Soding, 2005).

Brief Instructions

Following instructions should be followed in order to predict the 3D structure of the target protein (RASSF2) through HHpred.

- Open the web browser
- Go to HHpred homepage (http://toolkit.tuebingen.mpg.de/hhpred)
- An interface will open to submit the amino acid sequence of the target protein (RASSF2)
- Results will be sent through institutional mail ID depending upon the sequence (query) length and the number of queries sent to HHpred

Requirements

Input

HHpred required the FASTA sequence of the target protein (RASSF2) for structure prediction.

Sequence Submission

The amino acid sequence of the target protein (RASSF2) in FASTA format is submitted to HHpred (Fig. **30**) to build the 3D structure.

Fig. (30). HHpred sequence submission form.

Accessing Data

The user will have to wait according to the number of homologous sequences, the frequency and length of insertions and deletions for a prediction to complete. The summary information and the predicted structure will send to the user *via* mail.

Results Interpretation (Output)

Results are sent by e-mail. The PDB file can be downloaded from the e-mail.

Conclusion

IntFOLD2 generated the 3D models for RASSF2.

M4T

M4T (Multiple Mapping Method with Multiple Templates) is a comparative protein structure modeling server that uses a number of multiple templates to predict the model of the target protein (Struct *et al.,* 2009).

Introduction

M4t is considered as one of the excellent tools for comparing, coloring, annotating and mutating the 3D structures of the proteins. It predicts the structures on the concept of comparative modeling by using the combination of multiple templates and iterative optimization of alternative alignments.

Brief Instructions

Following instructions should be followed in order to predict the 3D structure of the target protein (RASSF2).

- Open the web browser
- Open M4t homepage (http://manaslu.aecom.yu.edu/M4T/)
- Submit the amino acid sequence of the target protein (RASSF2)
- Provide the institutional e-mail ID (abc@uni.edu.pk).
- Results will be sent through institutional mail depending upon the sequence (query) length and number of queries sent

Requirements

Input

The only input required for M4t is the FASTA sequence of the target protein (RASSF2).

Sequence Submission

The FASTA sequence of RASSF2 submitted to M4t for structure prediction (Fig. **31**).

Fig. (31). M4t Sequence submission form.

Accessing Data

The user will have to wait according to the number of homologous sequences, the frequency and length of insertions and deletions for a prediction to complete.

Results Interpretation (Output)

The results of the predicted structures through M4t contain the 3D model of the target protein (RASSF2), alignment, and energy profile. The model can be downloaded by clicking on the option "Download Model" (Fig. **32**).

M4T Server ver. 3.0
Comparative Modelling using a combination of multiple templates and iterative optimization of alternative alignments

M4T Message

Region of the target sequence covered by the model : 189 261

Template(s) used : 3ddc_57_132_B

DOPE score(*) : -14258.629883

Prosa2003 Zscore(*) " -6.15

Download alignment

Download model

Download Prosa2003 Energy Profile (raw data)

Download Prosa2003 Energy Profile (jpg format)

(*) For more details, read REMARK annotations in the model file

Please cite:
Bioinformatics. (2007) 23, 2558-65
Nucleic Acids Res. (2007) 35, W363-68
J. Struct. Funct. Genomics. (2009) 10,95-9

Fig. (32). Result Screen containing predicted model provided by M4t.

Conclusion

M4t predicted the model for RASSF2 and also aligned the target sequence with the template sequence.

MODELLER

The above-discussed tools are web-based modeling tools while MODELLER (a desktop based structure modeling tool) can be used to predict the 3D structure of a protein by user-defined parameters. MODELLER provides many flexible parameter lists that a user can utilize to produce the 3D structure of the protein manually.

Introduction

MODELLER is used for homology modeling to predict the 3D structures of the target protein (Eswar *et al.,* 2006; Marti-Renom *et al.,* 2000). The user provides an alignment of a sequence to be modeled with known related structures and MODELLER automatically calculates a model containing all non-hydrogen atoms. MODELLER implements comparative protein structure modeling by satisfaction of spatial restraints (Sali and Blundell, 1993) and can perform many additional tasks, including *de novo* modeling of loops in protein structures, optimization of various models of the protein structure with respect to a flexibly defined objective function, multiple alignment of the protein sequences and/or structures, clustering, searching of sequence databases and comparison of protein structures. MODELLER is considered as one of the most reliable and cited software for homology modeling. It predicts the authenticated loops, beta sheets, coils, helices and phi and psi angles by utilizing the template structure. This method generates valid and reliable structures by using suitable template having appropriate amino acids identity. Here, we will only predict the 3D structure of the target protein through one template.

Downloading and Installation

Initially, we have to download the suitable version of MODELLER (Fig. **33**) with our operating system and install the MODELLER .

Download & Installation

MODELLER is available free of change to academic non-profit institutions: you will, however, need to register for a license in order to use the software. It is also available through Accelrys for government research labs and commercial entities.

Modeller 9.15, released May 19th, 2015

To install MODELLER on this machine, we recommend the **Windows** package.

Windows (32-bit)	[GPG signature] Installation guide
Windows (64-bit)	[GPG signature] Installation guide
Mac (32-bit or 64-bit Intel)	[GPG signature] Installation guide
Linux (32-bit RPM)	Installation guide
Linux (64-bit x86_64 RPM)	Installation guide
Linux (32-bit Debian/Ubuntu package)	[GPG signature] Installation guide
Linux (64-bit x86_64 Debian/Ubuntu package)	[GPG signature] Installation guide
Generic Unix tarball	[GPG signature] Installation guide

Fig. (33). Packages for MODELLER.

As in the start of this book we mentioned to use RASSF2 sequence for all the analyses, we retrieved the amino acid sequence of RASSF2 from UniProt to predict the structure through MODELLER. In the BLASTp search, the corresponding protein had a higher similarity to the Crystal structure of NORE1A in complex with RAS.

Modeling Steps

The individual modeling steps of this example are explained below:

Searching for Structures Related to RASSF2

Firstly, it is necessary to generate a PIR format readable file of RASSF2 sequence for MODELLER (file "RASSF2.ali") (Fig. **34**).

```
>P1;RASSF2
sequence:RASSF2: 1 :A :326 : : : : :
MDYSHQTSLVPCGQDKYISKNELLLHLKTYNLYYEGQNLQLRHREEEDEFIVEGLLNISW
GLRRPIRLQMQDDNERIRPPPSSSSWHSGCNLGAQGTTLKPLTVPKVQISEVDAPPEGDQ
MPSSTDSRGLKPLQEDTPQLMRTRSDVGVRRRGNVRTPSDQRRIRRHRFSINGHFYNHKT
SVFTPAYGSVTNVRINSTMTTPQVLKLLLNKFKIENSAEEFALYVVHTSGEKQKLKATDY
PLIARILQGPCEQISKVFLMEKDQVEEVTYDVAQYIKFEMPVLKSFIQKLQEEEDREVKK
LMRKYTVLRLMIRQRLEEIAETPATI*
```

Fig. (34). Alignment file for MODELLER.

The first line contains the sequence code, in the format "*>P1; the name of protein*". The second line with ten (10) fields separated by nine (09) colons generally contains information about the structure file. The structure of the alignment file should be readable for MODELLER. If the given alignment sequence is of query, then the first field must be '*sequence*', second field must contain name, similar to that of '*the name of protein*' given in the first line of alignment field, third filed must contain '*starting sequence number*' and fourth field must have '*ending sequence number*', while all other fields could be kept empty. Only two of these fields are used for sequences, "*sequence*" (indicating that the file contains a sequence without known structure) and "*RASSF2.ali*" (the model file name). The rest of the file contains the sequence of RASSF2, with "*" marking its end. The standard one-letter amino acid codes are used. (Amino acids must be upper case; some lower case letters are used for non-standard residues).

BLAST (Basic Local Alignment Searching Tool)

Basic Local Alignment Search Tool (BLAST) is an algorithm for comparing primary biological sequence information, such as the amino acid sequences of different proteins or the nucleotides of DNA sequences. A BLAST search enables to compare a query sequence with a library or database of sequences, and identify library sequences that resemble the query sequence above a certain threshold (Fig. **35**) (Altschul *et al.*, 1997). There are different databases available that can be used as library against the query sequence and varies due to the nature of the protein and experiment. A database is required to be selected before performing BLAST search. For homology modeling experiments, it is better to select the "PDB" database. Alignment of query sequence against PDB would provide all those templates whose structures have been experimentally resolved. Here, for RASSF2 structure prediction, we utilized the PDB database for the template searching.

Fig. (35). Running BLAST for the respective sequence. 1) This is a text box where you can either provide fasta sequence, simple sequence, or you can even only provide uniProt ID, the blast program will automatically identify the type of input and get itself prepare for next process. 2) Here where you need to select the database for your search. In homology modeling experiments, it is always required to select 'Protein Data Bank (PDB)' database. PDB database is required because you need a template (whose structure is experimentally determined) against your query. 3) You can provide organism name here to restrict your search against given organism name. 4) There are a number of algorithms for search.

The FASTA sequence for RASSF2 is pasted to the interface of BLASTp by accessing its homepage. You can also give the accession number of the protein and BLASTp will get the protein sequence for analyses.

The BLAST search showed series of suitable templates for selection (Fig. **36**). A suitable template should have query coverage >50% and identity >55% for a reliable structure. For a selection of a suitable template, the parameters must be observed as E-value, query coverage, max score and identity.

Template PDB File

The PDB file of the selected template (3DDC) has to download from PDB (Fig. **37**) and have to place the PDB file in the same directory with script files otherwise we have to call the file with coding that will be difficult for the beginner.

Fig. (36). Different templates for RASSF2.

Fig. (37). Getting fasta sequence and PDB file of template.

Aligning RASSF2 with the Template

A good way of aligning the sequence of RASSF2 with the structure of **3DDC** is the **align2d.py** command in MODELLER. Although **align2d.py** is based on a dynamic programming algorithm, it is different from standard sequence-sequence alignment methods because it takes into account the structural information from the template when constructing an alignment. This task is achieved through a variable gap penalty function that tends to place gaps in solvent exposed and curved regions, outside secondary structure segments, and between two positions that are close in space. As a result, the alignment errors are reduced by approximately one-third relative to those that occur with standard sequence alignment techniques. This improvement becomes more important as the similarity between the sequences decreases and the number of gaps increases. Here, we want to mention one more thing that user can use other alignment software (ClustalW) for the alignment and give the alignment file to

MODELLER. The MODELLER script (Fig. **38**) aligns the RASSF2 sequence in file "RASSF2.ali" with **3DDC** structure in the PDB file "3DDC.pdb" (file "align2d.py").

```
env = environ()
aln = alignment(env)
mdl = model(env, file='3DDC')
aln.append_model(mdl, align_codes='3DDC', atom_files='3DDC.pdb')
aln.append(file='RASSF2.ali', align_codes='RASSF2')
aln.align2d()
aln.write(file='RASSF2-3DDC.ali', alignment_format='PIR')
aln.write(file='RASSF2-3DDC.pap', alignment_format='PAP')
```

Fig. (38). script for aligning the template with the target.

In this script, we created an empty alignment 'aln', and a new protein model 'mdl', into which read the chain A segment of the 3DDC PDB structure file. The **append_model()** command transfers the PDB sequence of the model to the alignment and assigns it the name of "**3DDC**" (**align codes**). The "**RASSF2**" sequence from the file "RASSF2.seq" was added to the alignment, using **append ()** command. The **align2d ()** command was executed to align the two sequences. The alignment was written out in two formats, PIR ("RASSF2-3DDC.ali") and PAP ("RASSF2-3DDC.pap"). The PIR format was used by MODELLER in the subsequent model building stage, while the PAP alignment format is easier to inspect visually. In the PAP format, all identical positions are marked with a "*" (file "RASSF2-3DDC.pap").

After preparing the input files for alignment the between RASSF2 sequence and 3DDC structure, we have opened the MODELLER command prompt from the start menu. MODELLER opens a black screen where we can run the commands for executing the alignment python scripts. We write the Mod from MODELLER and its version (*e.g* **mod9.14**) and then execute the python script (*i.e.* **align2d.py**) (Fig. **39**).

Model Building

Once a target-template alignment was constructed, MODELLER calculates a 3D model of the target completely automatically by using its **auto model** class. The script (Fig. **40**) will generate ten models of RASSF2 based on the 3DDC template structure and the alignment in file "RASSF2-3DDC.ali" (file "get-model.py").

Here, we just want to precisely explain the function of used classes of MODELLER in one line. The first line loads in the **automodel** class and prepares it for use. Create an **automodel** object, call it 'a', and set parameters to guide the model building procedure. **Alnfile** names the file that contains the target-template alignment in the PIR format. **Knowns** defines the known template structure(s) in **alnfile** ("RASSF2-3DDC.ali"). **Sequence** defines the name of the target sequence in **alnfile**. **assess_methods** request one or more assessment scores.

```
cx  Modeller
You can find many useful example scripts in the
examples\automodel directory.
It is recommended that you use Python to run
Modeller scripts. However, if you don't have Python installed,
you can type 'mod9.14' to run them instead.

E:\Research\Modeller9.14>mode.14 align2d.py_
```

Fig. (39). MODELLER command prompt showing the command for executing the python script for alignment.

```
# Step 4: model building
#
# This script should produce two models, Ifdx_my.B99990001.pdb and
# 1fdx_my.B99990002.pdb.

from modeller import *
from modeller.automodel import *      # Load the automodel class

log.verbose()
env = environ(rand_seed=-12312)  # To get different models from another script

# directories for input atom files
env.io.atom_files_directory = ['../atom_files']

a = automodel(env,
alnfile='RASSF2-3DDC.ali',        # alignment filename
knowns=('3DDC'),      # codes of the templates
sequence='RASSF2',            # code of the target
assess_methods=assess.GA341)  # request GA341 assessment
a.starting_model= 1              # index of the first model
a.ending_model  = 10             # index of the last model
# (determines how many models to calculate)
a.deviation = 4.0                     # has to >0 if more than 1 model

a.make()                              # do homology modelling
```

Fig. (40). Python script for model generation through MODELLER.

Starting_model and **ending_model** define the number of the models that are calculated. The user can increase and decrease the total number of models by changing the figures from starting_model and ending_model according to the nature of the experiment. The last line in the file calls the **make** method that actually calculates the models. We will again execute the python script for model generations in MODELLER command prompt window as depicted in Fig. (**41**).

Fig. (41). MODELLER Command prompt showing the get-model python script for model generation.

There will be many output files but the most important output file is "get-model.log" that contains warnings, errors, and other useful information including the input restraints used for modeling that remain violated in the final model. The MODELLER will assign molpdf values to all the predicted models (Fig. **42**) and the model will have the lowest molpdf value will consider the most reliable one according to MODELLER analyses. Here, we want to inform the user that it is not the final criteria to select the final reliable model and in next chapter, we will discuss the evaluation criteria of the predicted structures.

LIMITATIONS OF STRUCTURE PREDICTION TOOLS

Phyre2

- If the homology cannot be detected between a user-supplied sequence and a sequence of known structure, then modeling will either be impossible or extremely unreliable. This reflects the wider ongoing difficulty of the protein-folding problem. There are still no reliable methods to predict a protein structure purely from sequence alone without reference to known structures.
- It has the functionality to predict the phenotypic effect of a point mutation, but it

is unable to accurately determine, beyond the estimated position of a side chain, the wider structural effect of a point mutation.

```
>> Summary of successfully produced models:
Filename                               molpdf    GA341 score
-----------------------------------------------------------
RASSF2 B99990001.pdb              3670.30884       0.06946
RASSF2 B99990002.pdb              4316.43652       0.04175
RASSF2 B99990003.pdb              3955.36597       0.07801
RASSF2 B99990004.pdb              3897.07861       0.05255
RASSF2 B99990005.pdb              3312.38794       0.02493
RASSF2 B99990006.pdb              3704.33032       0.04622
RASSF2 B99990007.pdb              3912.39746       0.04721
RASSF2 B99990008.pdb              3716.84253       0.02740
RASSF2 B99990009.pdb              3800.79175       0.05289
RASSF2 B99990010.pdb              3806.58545       0.04292
```

Fig. (42). Models predicted by MODELLER.

I-TASSER

- I-TASSER server only accepts the protein sequence of the length up to 1,500 amino acids
- The protein modeling through I-Tasser could be performed by either academic user and commercially.
- It mostly takes 48 to 72 hours to model the protein structures.

RaptorX

- It is an automated web modeling that automatically identifies and utilize template, does not involve the manual evaluation.
- Sometimes, even after the completion of modeling, we didn't find the results through email due to an error in the server.
- Interactive visualization of RaptorX results requires Java and HTML5.

SWISS-MODEL

- The modeling results are non-experimental and must be considered with care as models may contain significant errors.
- All the information on the SWISS-MODEL site is provided "AS-IS", without any warranty, expressed or implied.
- Swiss-Model must not be used on any data which must remain confidential for legal or IP reasons.
- It generates the structure of the protein as it gets a template for a specific region and models that regions only while skips the non-aligned residues.
- The protein models generated from Swiss Model are not according to the given

sequence length.

3D-JIGSAW

- It fails when the significant homolog template is not present.
- It requires Ramol to visualize the 3D predicted models.

IntFOLD2

- It predicts the best model that may be not the best model for functional and other structural analysis.

M4t

- Occasionally, M4T may fail to provide a prediction. The main reason is usually that PSI-BLAST fails to find the homologous protein structure to the sequence. But even if PSI-BLAST succeeds to detect possible template(s), after running the MT module none of the PDB hits might be found to be suitable to model the target sequence.

MODELLER

- It cannot correct a wrong or suboptimal target-template alignment. The errors in the alignment will directly translate into errors in the model structures. The alignment errors are the leading source of the errors for homology models when the sequence similarity is below ~50%.
- The model is, on average, not more similar to the real target structure than the template used. (It means that normally, the improvements which the homology modelling software achieves do not fully compensate the alignment errors, on average). For this reason, the other limiting factor in model quality, besides the alignment quality, is the selection of the best template. This is a very difficult task for which a large number of methods and servers have been developed.
- Side chain placement gets unreliable below the sequence identities of 70%, simply because the position of the side chains is conserved less and less below this degree of similarity.
- Loops of more than three or four residues for which no corresponding template atoms are available cannot be reliably estimated. They are more or less guessed by the programs. There exist specialized loop modelling methods, but they are very time-consuming and cannot be offered over the web and the loops of more than 8 to 12 residues cannot be modelled reliably.

Table 4. Guideline for Structure Prediction.

Tool Name	Availability	Description	Work Flow	URL	References
Phyre2	Automated web-based modeling tool	Predicts 3d structure by using the principles and techniques of homology modeling	+Sequence submission =Predicted structures	http://www.sbg.bio.ic.ac.uk/phyre2/html/page.cgi?id=index	Lawrence *et al.*, 2011
I-TASSER	Automated web-based modeling tool	It employs threading based approach to build the 3d structures of proteins.	+Sequence *Folds recognition =Predict Models	http://zhanglab.ccmb.med.umich.edu/ I-TASSER/	Zhang *et al.*, 2010
RaptorX	Automated web-based modeling tool	It predicts secondary and tertiary structures even in the absence of closed homologs.	+FASTA sequence =Predicted Models	http://raptorx.uchicago.edu/Structure Prediction/predict/	Peng & Xu, 2011
SWISS-MODEL	Automated web-based modeling tool	It is an automated server to generate protein models based on homology modeling.	+Query Sequence =Results interpretations	http://swissmodel.expasy.org/ interactive	Biasini *et al.*, 2014
3D-JIGSAW	Automated web-based modeling tool	It builds 3d models for proteins based on homologues templates.	+Sequence =Results through Email	https://bmm.crick.ac.uk/~3djigsaw/	Tramontan *et al.*, 2001
IntFOLD2	Automated web-based modeling tool	It is integrated Protein Structure and Function Prediction Server.	+Sequence =Predict Models =Assess quality	http://www.reading.ac.uk/bioinf/IntFOLD/IntFOLD2_form.html	McGuffin *et al.*, 2012
HHpred	Automated web-based modeling tool	It is a sensitive homologs detection and structure prediction tool works on HMM-HMM-comparison.	+FASTA Sequence =Models	https://toolkit.tuebingen.mpg.de/hhpred	Soding, 2005
M4t	Automated web-based modeling tool	It utilizes multiple templates to predict the protein model.	+Query =Results =Alignment optimizations	http://manaslu.fiserlab.org/M4T/	Struct *et al.*, 2009
Modeller	Desktop based modeling Tool	It implements comparative modeling by satisfaction of spatial restraints.	+Sequence +Templates +Alignment =Models	https://salilab.org/modeller/download_installation.html	Eswar *et al.*, 2007

Protein 3D structure prediction approach employs various computational tools and webservers based on different methods. Computational structure prediction tools utilized in this chapter are summarized in Table **4**.

REFERENCES

Buenavista, M.T., Roche, D.B., McGuffin, L.J. (2012). Improvement of 3D protein models using multiple templates guided by single-template model quality assessment. *Bioinformatics, 28*(14), 1851-1857. [http://dx.doi.org/10.1093/bioinformatics/bts292] [PMID: 22592378]

Eswar, N., Webb, B., Marti-Renom, M.A., Madhusudhan, M.S., Eramian, D., Shen, M.Y., Pieper, U., Sali, A.

(2007). Comparative protein structure modeling using MODELLER. *Curr. Protoc. Protein Sci., Chapter 2*(November), 9.
[http://dx.doi.org/10.1002/0471140864.ps0209s50] [PMID: 18429317]

Kelley, L.A., Sternberg, M.J. (2009). Protein structure prediction on the Web: a case study using the Phyre server. *Nat. Protoc., 4*(3), 363-371.https://doi.org/http://dx.doi.org/10.1038/nprot.2009.2
[http://dx.doi.org/10.1038/nprot.2009.2] [PMID: 19247286]

Kelley, L.A., Bennett-Lovsey, R., Herbert, A.F. (2011). *Phyre: Protein Homology/analogy Recognition Engine. Structural Bioinformatics Group..* London, UK: Imperial College.

Mount, D.W. (2000). *Bioinformatics - Sequence and Genome analysis.*

Peng, J., Xu, J. (2011). RaptorX: exploiting structure information for protein alignment by statistical inference. *Proteins, 79* (Suppl. 10), 161-171.
[http://dx.doi.org/10.1002/prot.23175] [PMID: 21987485]

Rykunov, D., Steinberger, E., Madrid-Aliste, C.J., Fiser, A. (2009). Improved scoring function for comparative modeling using the M4T method. *J. Struct. Funct. Genomics, 10*(1), 95-99.
[http://dx.doi.org/10.1007/s10969-008-9044-9] [PMID: 18985440]

Söding, J. (2005). Protein homology detection by HMM-HMM comparison. *Bioinformatics, 21*(7), 951-960.
[http://dx.doi.org/10.1093/bioinformatics/bti125] [PMID: 15531603]

Tramontano, A., Leplae, R., Morea, V. (2001). Analysis and assessment of comparative modeling predictions in CASP4. *Proteins, 45* (Suppl. 5), 22-38.
[http://dx.doi.org/10.1002/prot.10015] [PMID: 11835479]

Zhang, Y., Murzin, A., Bateman, A., Ginalski, K., Rychlewski, L., Baker, D. (2008). I-TASSER server for protein 3D structure prediction. *BMC Bioinformatics, 9*(1), 40.
[http://dx.doi.org/10.1186/1471-2105-9-40] [PMID: 18215316]

## CHAPTER 5

Structure Evaluation

Abstract: The computational prediction of the protein 3D structures involves the mathematical modeling techniques based on statistical terms. That is why the resultant structures may have some structural issues related to β-sheets, α-helices, non-structural loops, amino acid chain angles and backbone psi, phi bond angles that pose potential energy constraints in the structure. The validation of the computationally predicted structure is mandatory that compares different parameters of the protein 3D structure with that of experimentally available protein structures. Here, in this chapter we will discuss the evaluation procedure of the 3D structures for their accuracy and physio-chemistry by numerous available tools.

Keywords: Anolea, ERRAT, Molprobity, PDBsum, Rampage, Structure evaluation.

MOTIVATION

The structures resolved through X-ray crystallography and NMR techniques provide better resolution, high accuracy, and structural information of a protein. Computational techniques use the mathematical models and algorithms to solve the protein structures where different tools implement the different algorithms. So, the accuracy of the model through computational methods need evaluation in terms of energy, stabilization, the angles, amino acid conformation at a specific place and other structural features. For these purposes, many tools are developed that evaluate the proteins on different terms. They compare the protein models with experimentally determined structures for their viability as an experimental model. This chapter introduces some most widely used protein evaluation tools and the interpretation of their results.

In this Chapter, we will follow the methodology of section 05 mentioned in Fig. (3).

MODEL EVALUATION TOOLS

The computationally predicted structures may have the possibility of the errors. The model evaluation (validation) checks the accuracy of the predicted model as by the representation of the real system. On the basis of experimentally designed structures, there are many tools to validate the models. The selection of the model is based on physiochemical properties of the proteins, the cell type of protein expression, and data from validation (Clarke *et al.,* 2000). Rampage, ERRAT molprobity and verify 3D are some of the most widely used model validation tools.

The recognition of the errors in experimental and theoretical models of the protein structures is considered as one of the major problems in structural bioinformatics. There is no single method that can consistently and accurately predicts the errors in the predicted structures. Various evaluation tools can be used for the assessment of the protein structure.

ERRAT

ERRAT evaluates the predicted models by analyzing the statistics of non-bonded interactions between different atom types.

Introduction

ERRAT uses the algorithm of protein structure verification that especially suited for evaluating the progress of crystallographic model building and refinement (Colovous *et al.,* 1993).

Brief Instructions

Following instructions should be followed in order to evaluate the 3D structure of the target protein (RASSF2).

- Open the web browser
- Go to ERRAT homepage (http://services.mbi.ucla.edu/ERRAT/)
- The coordinates should be submitted in PDB format
- Simply select it with the file dialog which activated by clicking on the browse button (choose file)
- Use the "Run ERRAT" option for the evaluation
- The result will be open in new window

Requirements

Input

The only input required for ERRAT is the PDB (file format) coordinates of the target protein (RASSF2).

Sequence Submission

The predicted structure of RASSF2 by utilizing different tools mentioned in the previous chapter was utilized for evaluation and the PBD file of RASSF2 was uploaded (Fig. **43**).

Upload a file: [Choose File] rassf2 i tasser.pdb

Please add these two numbers: 7 + 2 = [9]

[Run Errat]

Fig. (43). Giving input File (PDB) RASSF2 for ERRAT.

Results Interpretation (Output)

ERRAT generated a plot indicating the confidence and the overall quality of the model. A single output plot was given by the ERRAT which gives the value of the error function and position of the residue sliding window (Fig. **44**).

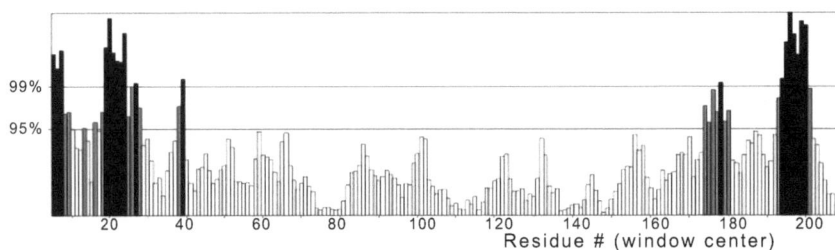

```
Program: ERRAT2
File: /var/www/SAVES/Jobs/4454959//errat.pdb
Chain#:1
Overall quality factor**: 82.075
```

*On the error ais, tow lines are drawn to indicate the confidence with which it is possible to reject regions that exceed that error value.
**Expressed as the percentage of the protein for which the calculated error value falls below the 95% rejection limit. Good hight resolution structures generally produce values around 95% or higher. For lower resolutions (2.5 to 3Å) the average overall quality factor is around 91%.

Fig. (44). ERRAT results for RASSF2.

- It shows an overall quality factor.
- On the error axis, two lines were drawn to indicate the confidence with which it is possible to reject that exceed the error value.
- Expressed as the percentage for which the calculated error value fall below 95% rejection limit. Good high-resolution structures generally produce values around

95% or higher. For lower-resolution, the average overall quality factor is around 91%. The structure having > 85% overall quality factor considered the reliable one and >90% considered excellent structure.

Conclusion

ERRAT evaluation tool was utilized for the evaluation of RASSF2 structure and 82.075% of an overall quality factor was observed.

RAMPAGE

Rampage is a model verification tool for the evaluation of predicted structures. It builds Ramachandran plot of the predicted protein 3D structures for analyses (Jeng *et al.,* 2011).

Introduction

Ramachandran plots show Φ and Ψ distributions of non-Glycine, non-Proline residues and give residues distribution. The phi and psi angles originated were plotted against each other to differentiate the favorable and unfavorable regions. These angles were used to evaluate the quality of regions.

Brief Instructions

Following instructions should be followed in order to evaluate the 3D predicted models of the target protein (RASSF2) through RAMPAGE.

• Open the web browser
• Go on RAMPAGE
• Upload PDB file of the target protein (RASSF2)
• Submit to the RAMPAGE
• A page will open which shows the Ramachandran plot

Requirements

Input

The input required for RAMPAGE is the PDB (file format) coordinates of the target protein (RASSF2).

Sequence Submission

The predicted model of RASSF2 was uploaded to RAMPAGE (Fig. **45**) for evaluation.

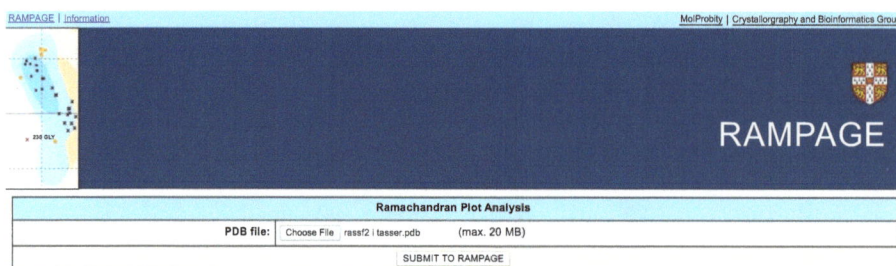

Fig. (45). RAMPAGE home.

Results Interpretation

The ramachandran plot of the target protein showed by rampage and also showed which residues are in the allowed region, favored region and in outliers. The more residues in the outlier, the more unreliable structure will be (Fig. **46**). The residues in total from the favored region and allowed region should be more than 95% for the reliable structure. The outliers can manually be corrected by different software including wincoot.

Number of residues in favoured region	(~98.0% expected)	:	216 (66.7%)
Number of residues in allowed region	(~2.0% expected)	:	69 (21.3%)
Number of residues in outlier region		:	39 (12.0%)

Fig. (46). RAMPAGE results for RASSF2.

Conclusion

RAMPAGE evaluated the predicted structure of RASSF2 and showed only 12 residues out of 326 amino acids in the outlier region.

ANOLEA

ANOLEA (Atomic Non-Local Environment Assessment) evaluation server evaluated the structure by energy calculation and gives non-local normalized energy Z-score of the predicted model.

Introduction

ANOLEA is a server that performs energy calculations on a protein chain evaluating the non-local environment of each heavy atom in the molecule. The energy of each pairwise interaction in this non-local environment is taken from a distance dependent knowledge based mean force potential that has been derived from a database of 147 non-redundant protein chains with a sequence identity below 25% and solved by X-ray crystallography with a resolution lower than 3Å (Melo *et al.*, 1997).

Brief Instructions

Following instructions should be followed in order to evaluate the 3D structure of the target protein (RASSF2).

- Open the web browser
- Go to the ANOLEA homepage
- Upload the amino acid sequence of the target protein
- The coordinates should be submitted in PDB format
- Select the file dialog by clicking on the Browse option (Choose file)
- Click the send file option for submission (submit server)

Requirements

Input

The predicted structure of RASSF2 submitted to ANOLEA for evaluation (Fig. 47).

ANOLEA FORM

● **Job title (optional) :**

● **Window Average:** (Help)

5 amino acids ▾

● **Threshold to define a high energy amino acid :** 0.000   (Help)

● **Protein chain information (mandatory):** (Help)

Please, in the box below, write the sequence of all the chain identifiers of your protein. In first place put the chain on which you want to perform the calculations. The non-local energy profile will be performed only on the chain indicated by the first chain identifier of the sequence supplied in the box below, but considering all the atoms of the other chain identifiers If you are submiting a monomeric protein you must write the chain identifier if there is one. If there is not a chain identifier, please leave this box free. We strongly recommend to read the help provided for this issue when using *ANOLEA* for the first time.

Chain names:

● Indicate the name of the PDB file that contains the coordinates. Use this field to browse you file (recommended) or paste the ''ATOM'' section of your PDB file in the text area supplied below. This parameter is mansatory. Do not paste anything if you have indicated a filename in the box:

File Upload: Give the name of the PDB file in your system (it must be a text file not encoded I) | Choose File | No file chosen

Warning: All heteroatoms supplied will be removed and will not be considered in the calculations.

● Have you the CHIME plug-in installed ?: (Help)

No ▾

If you have not installed **MDL CHIME** on you computer, you can download it from here (MDL CHIME is free and we recommend to use it to analyze the results of the calulation). A complete information about MD CHIME capabilities is available here

MDL CHIME requires *Netscape 2.0 or later (depending on the platform you are using)*

● How do you want the output ?:

NO CHIME   ▾

Fig. (47). ANOLEA form for Sequence submission.

Results Interpretation

It tells following information about RASSF2 (Fig. **48**)

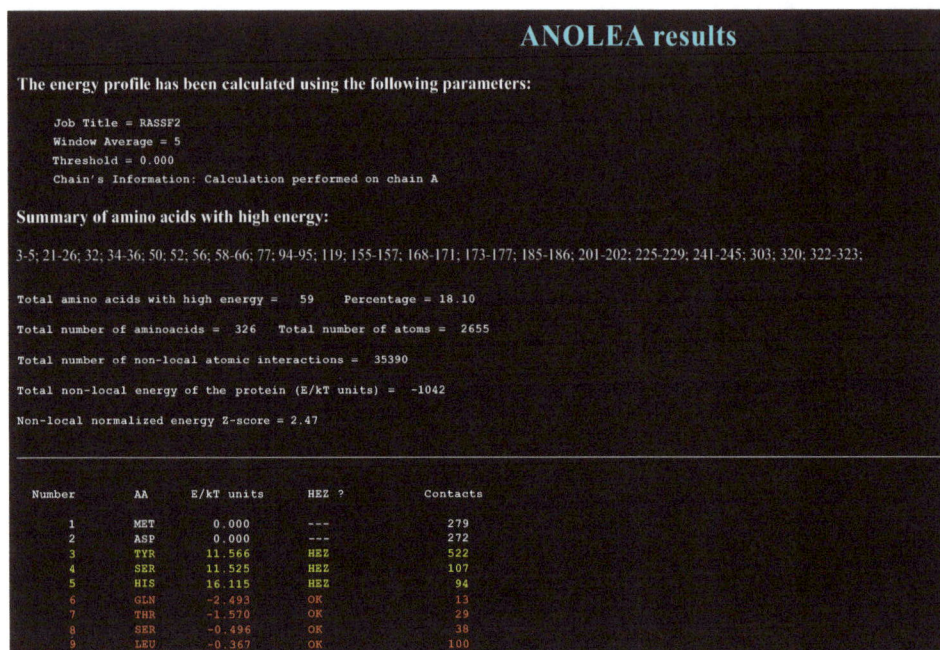

ANOLEA results

The energy profile has been calculated using the following parameters:

    Job Title = RASSF2
    Window Average = 5
    Threshold = 0.000
    Chain's Information: Calculation performed on chain A

Summary of amino acids with high energy:

3-5; 21-26; 32; 34-36; 50; 52; 56; 58-66; 77; 94-95; 119; 155-157; 168-171; 173-177; 185-186; 201-202; 225-229; 241-245; 303; 320; 322-323;

Total amino acids with high energy = 59   Percentage = 18.10

Total number of aminoacids = 326   Total number of atoms = 2655

Total number of non-local atomic interactions = 35390

Total non-local energy of the protein (E/kT units) = -1042

Non-local normalized energy Z-score = 2.47

Number	AA	E/kT units	HEZ ?	Contacts
1	MET	0.000	---	279
2	ASP	0.000	---	272
3	TYR	11.566	HEZ	522
4	SER	11.525	HEZ	107
5	HIS	16.115	HEZ	94
6	GLN	-2.493	OK	13
7	THR	-1.570	OK	29
8	SER	-0.496	OK	38
9	LEU	-0.367	OK	100

Fig. (48). ANOLEA tool showing the non-local atomic interactions and normalized z-scores.

- The summary of the amino acids with high energy
- The total amino acids with high energy and its percentage
- The total numbers of atoms
- The total numbers of non-local atomic interaction
- The total non-local energy of the protein (E/KT unit)
- The non-local normalized energy score

Conclusion

So, ANOLEA can be helpful to evaluate the 3D predicted structure. It determines the high energy of amino acids.

MOL PROBITY

Mol Probity is a structure-validation web service that provides a broad-spectrum solidly based evaluation of model quality at both the global and local levels for both proteins and nucleic acids (Chen *et al.,* 2009).

Introduction

It evaluates the predicted structure on the broad-spectrum solidly based evaluation of model quality at global and local levels for the proteins and nucleic acids. It is a freely available online tool for protein evaluation (Davis *et al.,* 2007).

Brief Instructions

Following instructions should be followed in order to evaluate the 3D structure of the RASSF2 protein through Mol probity online tool.

- Open web browser
- Go to Mol Probity homepage
- Upload the .pdb file of RASSF2
- Interpret the results

Requirements

Input

The input required for MolProbity is the PDB (file format) coordinates of the target protein (RASSF2) (Fig. **49**).

Fig. (49). MolProbity Submission form.

- The main page opens in which different options were given (Figs. **50** and **51**)

- Click on the analyze geometry without all atom contacts
- The next page opens in which different output options were given
- Choose the option according to the nature of the experiment and protein, we choose default option here
- Click on run programs to perform the analysis
- Output geometry of RASSF2 displayed for analyses (Fig. **52**).

Fig. (50). Instructions of MolProbity Sequence Submission Form.

Fig. (51). Instructions of MolProbity Sequence Submission Form.

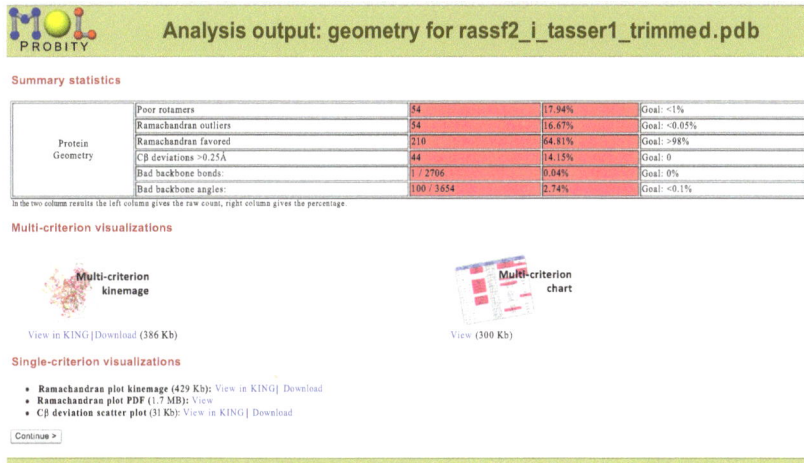

Fig. (52). MolProbity results for RASSF2.

Results Interpretation

- Results evaluated by MolProbity showed the Ramachandran outliers, favored region, Cβ deviations, bad backbone angles and bonds (Fig. **52**).
- Multi criterion chart showed detailed information of each and every residue of the target protein (Fig. **53**).

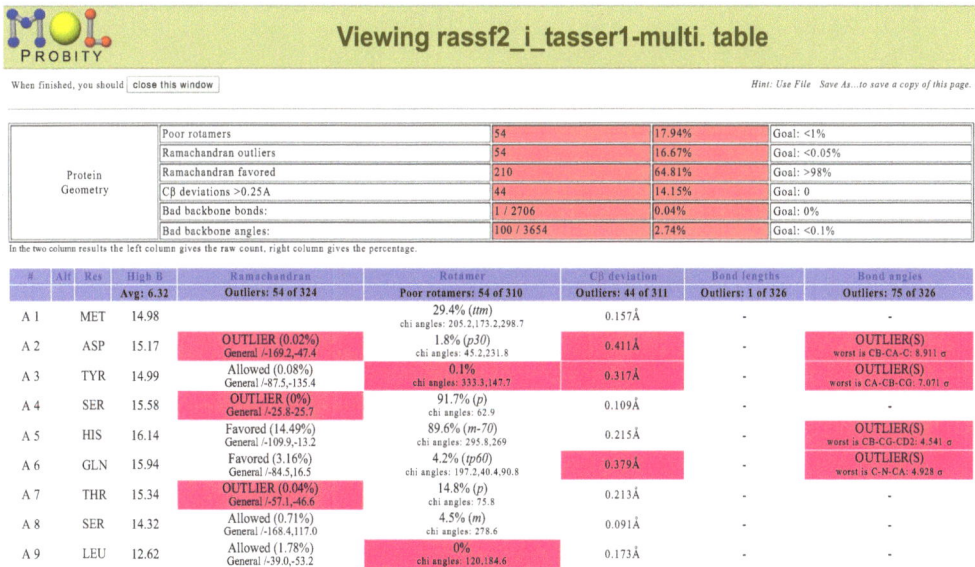

Fig. (53). Molprobity results for RASSF2.

Conclusion

MolProbity showed the poor rotamers, outliers and deviated angles of the predicted structure. The poor rotamers, deviated angles and outliers can be corrected by different tools to make the structure more reliable.

VERIFY 3D

Verify3D also evaluates the predicted structures.

Introduction

It determines the compatibility of an atomic model with its own amino acids sequences ID by assigning a structured class based on its location and environment (alpha, beta, loop, polar, non-polar). It also compares the results with the library of the reliable experimentally predicted structures (Bowie *et al.*, 1997).

Brief Instructions

Following instructions should be followed in order to evaluate the 3D structure of the target protein.

- Open web browser
- Go to VERIFY 3D homepage
- The coordinates should be submitted in PDB format
- Select the file dialog by clicking the choose file option
- Click the send file option to run VERIFY 3D
- The result will be open in new window

Requirements

Input

Verify 3D required the predicted structure in PDB (file format) coordinates of the target protein (RASSF2).

Sequence Submission

The predicted structure of RASSF2 was uploaded to Verify 3D (Figs. **54a**, **54b**) for evaluation.

Fig. (54a). Evaluation of predicted 3D model by verify 3D.

Upload a file: [Choose File] rassf2 i tasser.pdb

Please add these two numbers: 1+ 10 = [11] _What's this?_

[Run Verify 3D]

Fig. (54b). Evaluation of Predicted 3D model by verify 3D.

Results Interpretation

Verify3D displayed the results in the form of a graph (Fig. **55**).

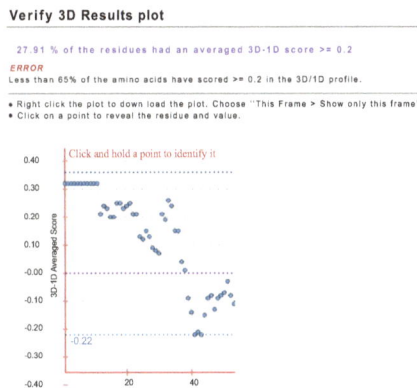

Fig. (55). Verify 3D graph for RASSF2.

Conclusion

The verify3D graph shows a 1D-3D comparison of the respective protein.

PDBSUM GENERATE

PDBsum generate provides Procheck plots to assess the stereochemical properties of the predicted structure. It is considered that in all given tools for structural assessment, PDBsum rather provides more accurate results.

Introduction

Procheck evaluates the normal and unusual geometry of the residues in the predicted structure and compared the stereochemical parameters derived from high-resolution refined structures (Laskowski *et al.,* 2001). Stereochemical parameters define as the atoms in all residue types, what are the atom pairs that are covalently bonded to each other and what are the ideal bond lengths, angles, dihedral angles, and improper dihedral angles.

Brief Instructions

Following instructions should be followed in order to evaluate the 3D structure of the target protein.

- First, open the browser
- Go to PDBsum
  (https://www.ebi.ac.uk/thornton-srv/databases/pdbsum/Generate.html)
- Choose the respective file to upload
- Institutional e-mail ID have to mention in to the appropriate section for the results

Requirements

Input

The input to PDBsum Generate is a single file containing the coordinates of the target protein structure (PDB file). The PDB file uploaded at the PDBsum page and the form allowed to upload PDB-format file and generated a full set of PDBsum structural analyses.

Sequence Submission

The PBD file of RASSF2 was uploaded to PDBsum Generate for evaluation (Fig. **56**).

Fig. (56). PDBsum Generate submission Form.

Accessing Data

The data of PDBsum have the procheck ramachandran plot stored in a secure password protected area. The PDBsum identifier will be in the form xNNN (x=letter, NNN=Sequential number). The results will send through the given mail ID.

Results Interpretation

By clicking on the link provided in the mail and the results will be observed for analyses.

By clicking on the procheck box at the right side, another window will open (Fig. 57).

Fig. (57). Procheck results provided by PDBsum generate.

Click on procheck option at the left side to get the plots for the submitted 3D model of the target protein (Fig. **58**). 10 different files or plots were appeared having stereochemical properties of the 3D model.

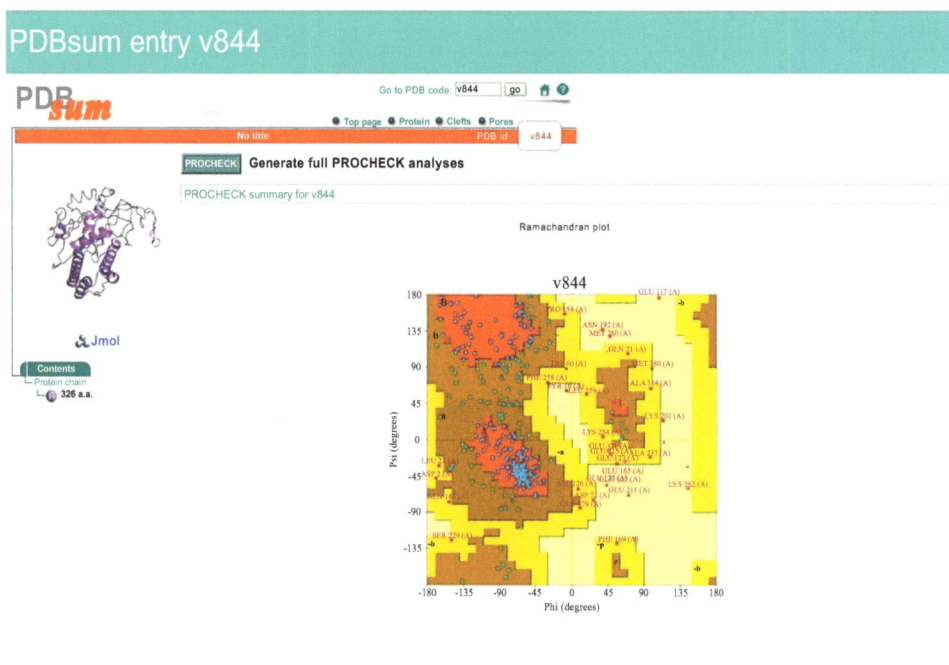

Fig. (58). Procheck full analysis.

Output

The outputs comprised a number of plots, together with a detailed residue-by-residue listing.

Customization of Plots

The plots produced by PROCHECK can be customized by amending the parameter file called procheck.prm. Following plots were obtained from the procheck analyses.

1. Ramachandran plot
2. Gly & Pro Ramachandran plots
3. Chi1-Chi2 plots
4. Main-chain parameters
5. Side-chain parameters
6. Residue properties
7. Main-chain bond length distributions

8. Main-chain bond angle distributions
9. RMS distances from planarity
10. Distorted geometry plots

Conclusion (Procheck Summary)

Residue properties (Fig. **59**)

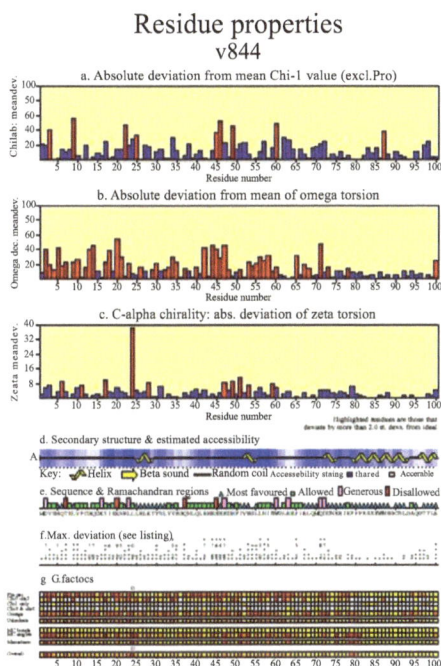

Fig. (59). Residue properties of RASSF2.

The residue number was given in the original coordinates file. The chain identifier picked up from the original coordinates file.

The sequential number was starting at 1 for the first residue and numbering the residues sequentially then. This may differ from the residue numbering given in the original coordinates file.

RAMACHANDRAN PLOT

The residues were displayed by a single letter code in Ramachandran plot and showed where the residue lies. The Ramachandran plot for the target protein (RASSF2) shown in Fig. (**60**).

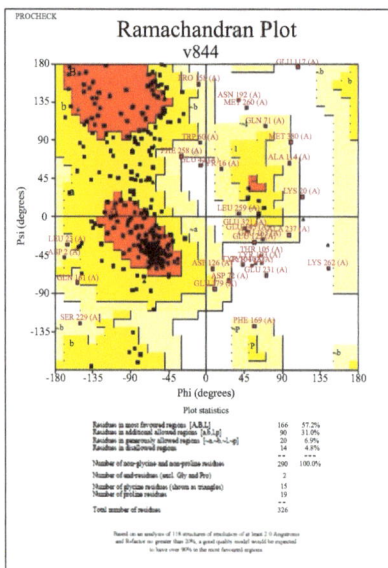

Fig. (60). Ramachandran Plot for RASSF2.

Chi-1 Dihedral Angle

Three separate columns were given for the three possible conformations of chi-1 (Fig. **61**).

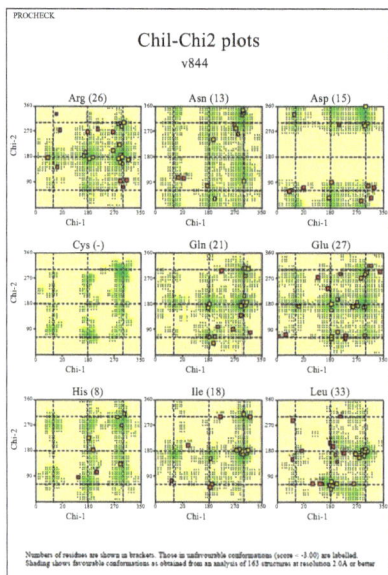

Fig. (61). Chi1-Chi2 Plots for RASSF2.

Chi-2 Dihedral Angle

The values for chi-2 dihedral angles in the **Trans** conformation and the shaded region showed the favored regions (Fig. **61**).

All Ramachandrans

Ramachandran plots for all the residue types were also generated and shaded region showed the favored region (Fig. **62**).

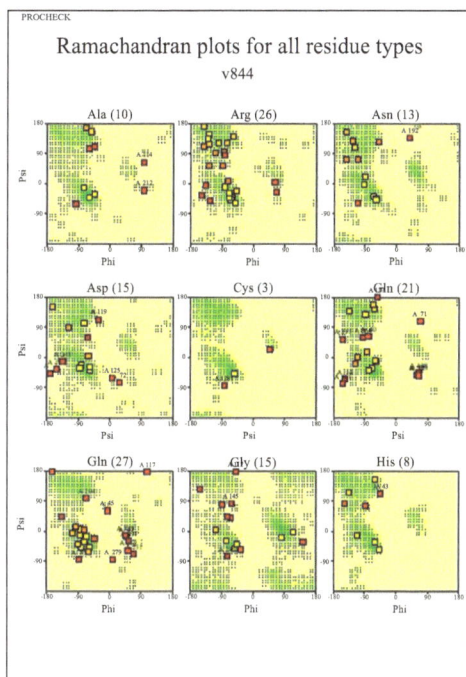

Fig. (62). Ramachandran Plot for all residues of RASSF2.

Main Chain Parameters

The main chain parameters including %-age residues in A, B, L, Omega angle st, Bad contacts/100 residues, Zeta angle st dev, H-bond energy st dev and Overall G-factor (Fig. **63**) of the predicted structure were also analyzed.

Side Chain Parameters

Chi-1 gauche minus st, Chi-1 Trans st dev, Chi-1 gauche plus st dev, Chi-1 pooled st dev, Chi-2 Trans st dev were the side chain parameters (Fig. **64**) for evaluating the predicted structure.

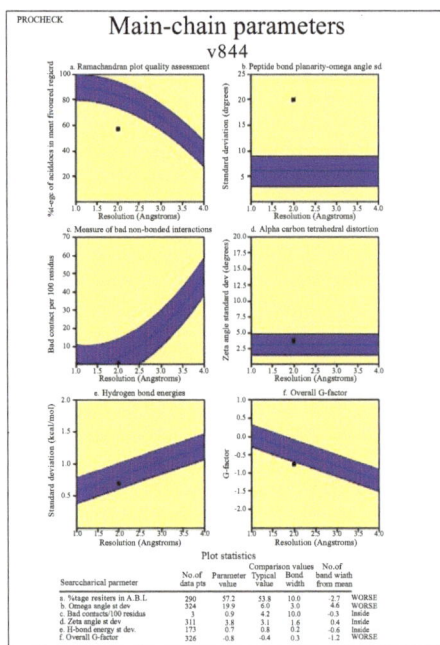

Fig. (63). Main chain Parameters for RASSF2.

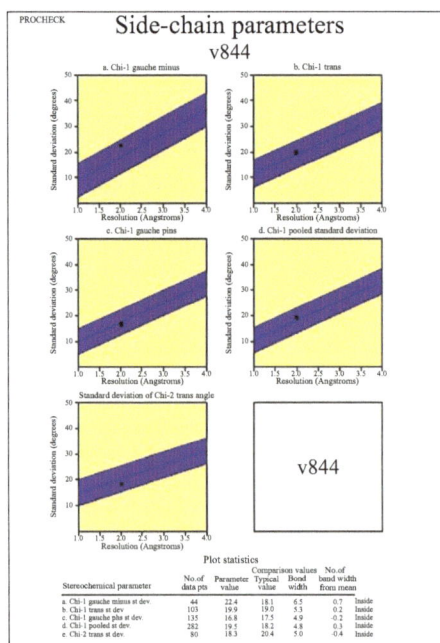

Fig. (64). Side Chain Parameters for RASSF2.

Main Chain Bond Lengths

C-N, C-O, Calpha-C, Calpha-Cbeta, N-Calpha bond lengths are included in the main chain bond length.

Main Chain Bond Angles

C-N-Calpha, Calpha-C-N, Calpha-C-O, Cbeta-Calpha-C, N-Calpha-C, N-Calpha-Cbeta, O-C-N bond angles are included in these parameters (Fig. **65**).

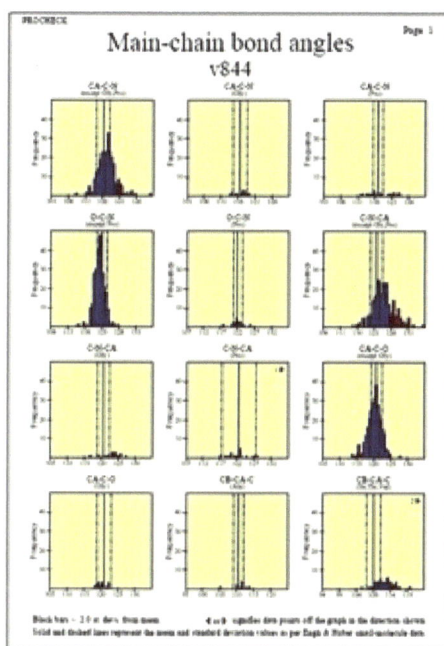

Fig. (65). Main chain Bond Angles of RASSF2.

Planar Groups

Histograms show RMS distances of planar atoms from the best-fit plane. Black bars indicate large deviations from planarity: RMS dist > 0.03 for rings, and > 0.02 otherwise (Fig. **66**).

LIMITATIONS OF STRUCTURE EVALUATION TOOLS

ERRAT

• It only calculates the non-bonded atomic interactions having distance more than 2.5 angstroms.

- It generates the multiple plots instead of a single plot for the protein having more than 300 amino acids.
- Sometimes, it fails to recognize the structure and its quality factor.

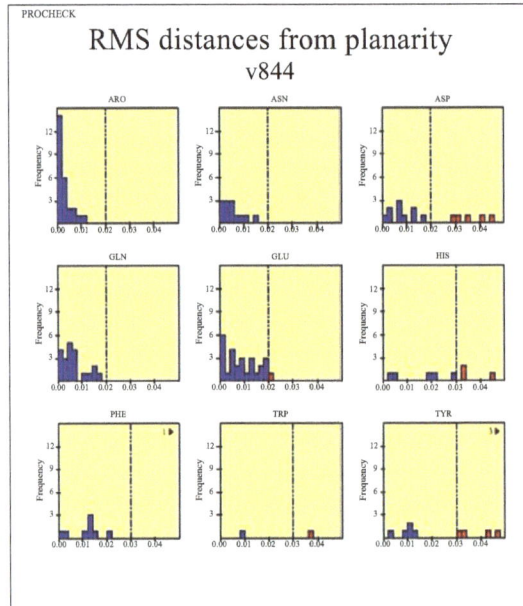

Fig. (66). Distances from Planarity.

RAMPAGE

- Sometimes, models having only loops give a good percentage of the favored residues.
- It only checks the phi-psi angles and torsional angles to evaluate the structures.

Verify3D

- The quality of difficult fold recognition models is below the standards of the benchmarks used to tune the majority of quality assessment methods, making application of such methods problematic in difficult cases.

PDBsum

- The server has to be contacted to (re)load a page when navigating through the pages, thereby slowing down the work.

Structure evaluation tools with their availability, features, methodology, web link and references are mentioned in Table **5**.

Table 5. Guideline for Structure Evaluation.

Tool Name	Availability	Description	Work Flow	URL	References
ERRAT	Web-based evaluation tool	It assesses the 3d model by analyzing the non-bonded atomic interactions.	+3d model =Quality Factor =Graph	http://services.mbi.ucla.edu/ERRAT/	Colovous & Yeates, 1993
RAMPAGE	Web-based evaluation tool	It generates ramachandran plot to evaluate protein model.	+Model =Plot =Residues distributions	http://mordred.bioc.cam.ac.uk /~rapper/rampage.php	Lovell *et al.*, 2002
ANOLEA	Web-based evaluation tool	It performs energy calculations and gives non-local normalized energy z-scores.	+Model *Energy calculations =Z-scores	http://protein.bio.puc.cl/anolea/	Melo *et al.*, 1997
Mol Probity	Web-based evaluation tool	It provides broad spectrum solidly based evaluation of model quality at global and local levels	+Structure =Protein Geometries analyses	http://molprobity.biochem.duke.edu/	Chen *et al.*, 2009
Verify 3D	Web-based evaluation tool	It determines the compatibility of an atomic model with its own amino acids sequences	+Model *3D-1D comparison =Graph	http://services.mbi.ucla.edu/Verify_3D/	Bowie *et al.*, 1997
PDBsum	Web-based evaluation tool	It evaluates the protein models through stereochemical properties	+Model *Ramachandran Plot +Stereochemical properties	https://www.ebi.ac.uk/thornton-srv/databases/pdbsum/Generate.html	Laskowski *et al.*, 2001

REFERENCES

Chen, V.B., Arendall, W.B., III, Headd, J.J., Keedy, D.A., Immormino, R.M., Kapral, G.J., Murray, L.W., Richardson, J.S., Richardson, D.C. (2010). MolProbity: all-atom structure validation for macromolecular crystallography. *Acta Crystallogr. D Biol. Crystallogr., 66*(Pt 1), 12-21.
[http://dx.doi.org/10.1107/S0907444909042073] [PMID: 20057044]

Colovos, C., Yeates, T.O. (1993). Verification of protein structures: patterns of nonbonded atomic interactions. *Protein Sci., 2*(9), 1511-1519.
[http://dx.doi.org/10.1002/pro.5560020916] [PMID: 8401235]

Davis, I.W., Leaver-Fay, A., Chen, V.B., Block, J.N., Kapral, G.J., Wang, X., Murray, L.W., Arendall, W.B., III, Snoeyink, J., Richardson, J.S., Richardson, D.C. (2007). MolProbity: all-atom contacts and structure validation for proteins and nucleic acids. *Nucleic Acids Res., 35*(Web Server issue) (Suppl. 2), W375-83.
[http://dx.doi.org/10.1093/nar/gkm216] [PMID: 17452350]

Eisenberg, D., Lüthy, R., Bowie, J.U. (1997). VERIFY3D: assessment of protein models with three-dimensional profiles. *Methods Enzymol., 277*, 396-404.
[http://dx.doi.org/10.1016/S0076-6879(97)77022-8] [PMID: 9379925]

Finkelstein, P.L., Ellestad, T.G., Clarke, J.F., Meyers, T.P., Schwede, D.B., Hebert, E.O., Neal, J.A. (2000). Ozone and sulfur dioxide dry deposition to forests: Observations and model evaluation. *J. Geophys. Res. D Atmospheres, 105*(D12), 15365-15377.
[http://dx.doi.org/10.1029/2000JD900185]

Jeng, W.Y., Wang, N.C., Lin, M.H., Lin, C.T., Liaw, Y.C., Chang, W.J., Liu, C.I., Liang, P.H., Wang, A.H. (2011). Structural and functional analysis of three β-glucosidases from bacterium Clostridium cellulovorans, fungus Trichoderma reesei and termite Neotermes koshunensis. *J. Struct. Biol., 173*(1), 46-56.
[http://dx.doi.org/10.1016/j.jsb.2010.07.008] [PMID: 20682343]

Laskowski, R.A. (2001). PDBsum: summaries and analyses of PDB structures. *Nucleic Acids Res., 29*(1), 221-222.
[http://dx.doi.org/10.1093/nar/29.1.221] [PMID: 11125097]

Lovell, S.C., Davis, I.W., Arendall, W.B., III, de Bakker, P.I., Word, J.M., Prisant, M.G., Richardson, J.S., Richardson, D.C. (2003). Structure validation by Calpha geometry: φ,ψ and Cbeta deviation. *Proteins, 50*(3), 437-450.
[http://dx.doi.org/10.1002/prot.10286] [PMID: 12557186]

Melo, F., Devos, D., Depiereux, E., Feytmans, E. (1997). ANOLEA: a www server to assess protein structures. *Int. Conf. Intel. Syst. Mol. Biol., 5*, 187-190.

Sehgal, S. A., Tahir, R. A., Shafique, S., Hassan, M., Rashid, S. (2014). Molecular modeling and docking analysis of CYP1A1 associated with head and neck cancer to explore its binding regions. *J Theor Comput Sci., 5*(112), 2.

Von Grotthuss, M., Pas, J., Wyrwicz, L., Ginalski, K., Rychlewski, L. (2003). Application of 3D-Jury, GRDB, and Verify3D in Fold Recognition. *Proteins: Structure, Function and Genetics, 53*, 418-423.
[http://dx.doi.org/10.1002/prot.10547]

CHAPTER 6

Visualization of Predicted Structure

Abstract: The structures of proteins generated experimentally or computationally are in a .pdb files and to visualize these structures are the key need of new era. The visualization of the predicted models helps to visualize including loops, α helices, β pleated sheets and turns for the better understanding of the structure. Numerous tools are available for visualizing the structures and also to edit or performing different functions as per the need of experiment. Here, we will focus on the basic visualizing functions by utilizing Chimera.

Keywords: Balls and stick model, Chimera, Energy minimization, Model visualization.

MOTIVATION

The protein structures are usually stored in .pdb file format. The visualization of .pdb files is necessary to learn and highlight important structural features. Protein structures can be visualized in the 3D space foregrounding atomic properties. There are many tools available that can help to visualize the proteins in 3D view. These tools also help to identify the domain/motif structures, the interactions of protein atoms with atoms of the interacting partners and to calculate many other related parameters including size of ligand pocket, the behavior of the interactions, angle and rotamers of the residues. This chapter will focus on UCSF Chimera, which is the most widely used tool to visualize the structure files. UCSF Chimera also has many built-in functions for performing different tasks on the structure files.

In this Chapter, we will follow the methodology of section 06 mentioned in Fig. (**3**).

UCSF CHIMERA

The primary programming language of Chimera is Python. Python is easy to learn and readable interpreted and object-oriented programming language. Chimera is divided into an extensions and cores. The higher level of functionalities is performed by extensions while the core provides basic services and molecular graphics capabilities. The Chimera has all the basic and advanced concepts of

Sheikh Arslan Sehgal, Rana Adnan Tahir, A. Hammad Mirza & Asif Mir

functionalities including molecules, atoms and bonds. Chimera is a good and accurate both in memory usage, speed and prefers programmability over performance.

As RASSF2 as an experimental protein in this work, we utilized the predicted model of RASSF2 for visualization through UCSF Chimera. We tried to reveal the basic and advance mandatory options of UCSF Chimera that is essentials for predicting active bio-molecules.

Introduction

UCSF Chimera is a highly extensible program for interactive visualization and analyses of the molecular structures and related data, including density maps, super molecular assemblies, sequence alignments, molecular docking results, trajectories, and conformational ensembles. The high-quality images and animations can be generated by using Chimera (Pettersen *et al.,* 2004).

Brief Instructions

Following instructions should be followed in order to visualize the 3D structure of the RASSF2.

- Open UCSF Chimera
- Open the PDB file as input file

Visualization Through Different Options

UCSF Chimera can open many file types including .pdb file and the 3D structure of the protein shown on the screen. Different functions can be performed on protein by using different options. UCSF Chimera has a menu bar with different options including File, Select, Action, Presets, and Tools. In this chapter, we will discuss these functions one by one.

File Option

In the file option, open, fetch by ID, restore the session, save session, save image, save PDB and save MOL file are the major options to do on (Fig. **67**).

Saving the Session Status

The Chimera session (the actual representation) can be saved for future use or modifications. This can be made by using the "**File/Save Session As...**" option from the menu bar. Select a file name and click the "**Save**" option. The saved file will be actually a python script and will have the ".py" extension.

Fig. (67). File Function.

Saving Image

The selected residues or any selected atom can be clear out and also to select any residue or atom by using "**Clear/Select Selection**" respectively. Choose an attractive and clear orientation and zoom out or zoom in provide a satisfying point of view. The saving of the generated image should be in good resolution and for that we selected the "**File/Save Image**" option from menu item. In the new window, click on "**Image Setup**" and select an image resolution of 300 pixels per unit. Click the "**Save**" option in this "**Preferences**" window. In the "**Save Image**" window, select "**Maintain current aspect ratio**", and enter an "**Image width**" of 6 inches. Click the "**Save As**". Wait while the image is calculated. Finally, choose "**File name**" and click the "**Save**" (Fig. **68**).

Fig. (68). Saving the image.

Select Option

There are few sub-options in Select option and we tried to discuss the significant sub-options below.

Structure

The secondary structure element of the RASSF2, *i.e.* strand, helix or loop can be shown. Go to the **"Select"**, **"Structure"** and **"secondary structure"** (Fig. **69**).

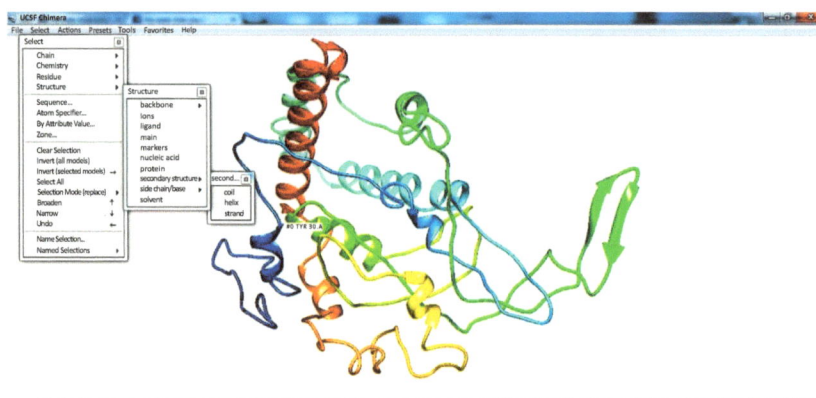

Fig. (69). Showing structure of RASSF2.

Residue

The residues of RASSF2 can be seen by **"Select"** and **"Residue"** and a new bar will open and the user can choose by their own choice of residues or according to the need of experiment (Fig. **70**).

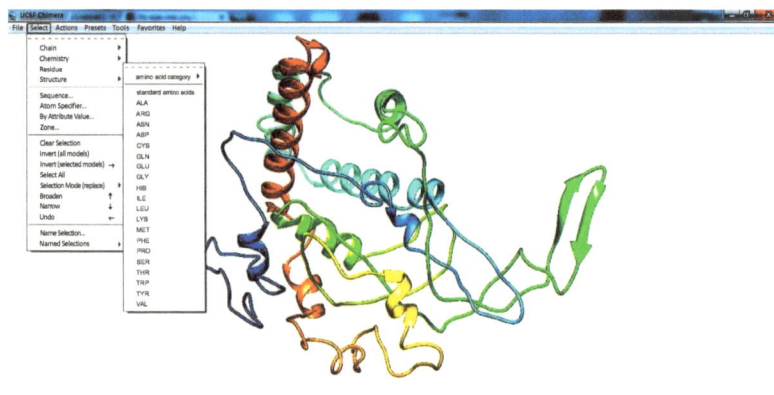

Fig. (70). Viewing the Residues of RASSF2.

Chemistry

In the Chemistry option, the user can analyze and visualize the elements, functional groups and atom type identity (IDATM type) can be found (Fig. **71**).

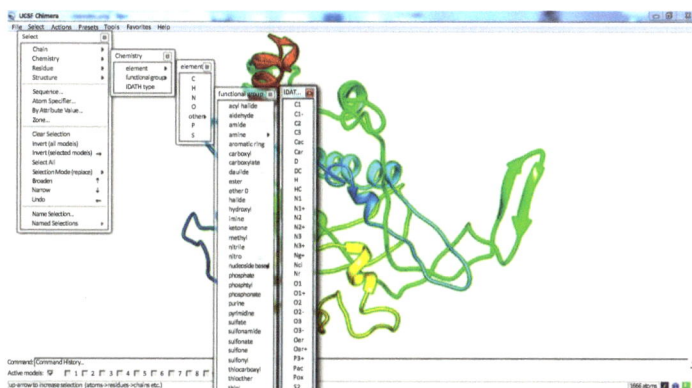

Fig. (71). Viewing IDATM type.

Action

Action option also has sub-options and the key options are discussed below.

Atoms/Bonds

The specific or all the atoms and bonds of the target protein can be seen by this option. Click on the "Atoms/Bonds" and "show option". The Atoms/Bonds can be represented to **"stick"**,**"ball and stick""sphere"** and **"wire"** styles and can also be modified (Fig. **72**).

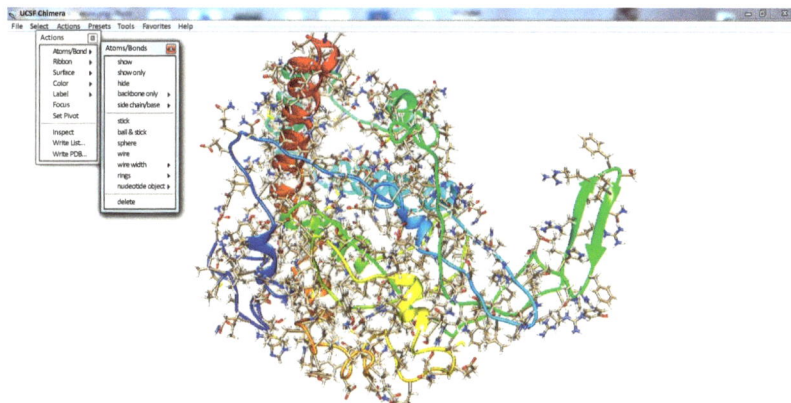

Fig. (72). Viewing the atoms and bonds of RASSF2.

Ribbon

The ribbon structure of the target protein can be used for attractive visualizations and also to modify the ribbons representation to "edge", "flat" and "rounded" shapes (Fig. **73**).

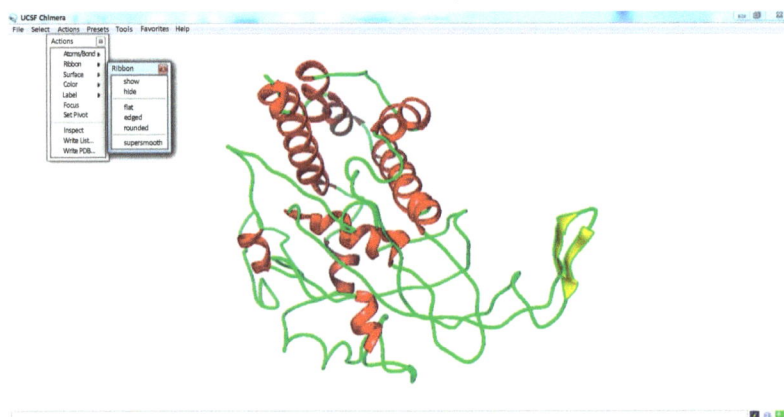

Fig. (73). Ribbon Representation of RASSF2.

Showing Surface

The surface of the target protein can also be selected by open the **"Surface"** window using **"Action/Surface"**. Click the **"show"** and modify the surface representation to **"mesh"** and **"dot"** as per need. The surface can be made transparent by using "**Surface**" window, click on"**transparency**" and select 80% or any transparency as per need (Fig. **74**).

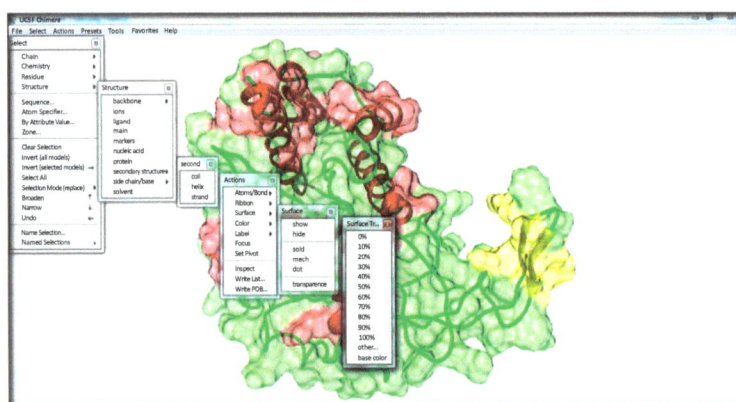

Fig. (74). Representing the surface of RASSF2.

Coloring the Secondary Structure Elements

The secondary structure elements, including strands, helix or loop can also be colored by click on the "**Select**" in the menu bar and select secondary structure in sub menu. Select the helix and choose the "**Actions**" menu item, and click on the color in sub-menu. Click on the "**All option**" and a new window will be opened. Choose the color and select.

Check the "**atoms/bonds**" box so that the color changing will be applied only to atoms and bonds. Select a color from the left column that will be used for all atoms. Another possibility is to color all atoms according to their atom types. This is obtained by clicking "**by element**" (Fig. **75**).

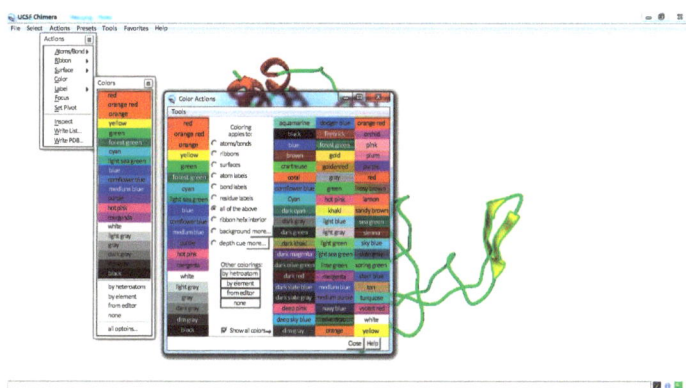

Fig. (75). Coloring the atoms, residues etc.

Label Residue

Open the "**Label**" window by choosing the "**Actions/Label**" menu item. In the "**Label**" window, choose "**Option**", "**Custom**" and click "**Ok**" (Fig. **76**). The user can also change the colors of the labels by using the "**Color**" window, checking "**residue labels**" and change the color.

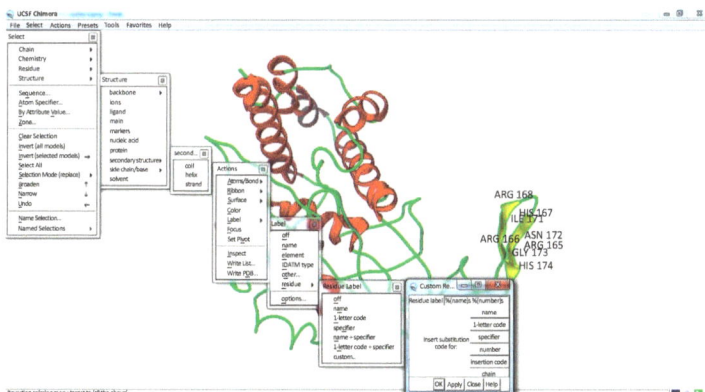

Fig. (76). Labeling the residues.

Presets Option

We can change and set the background and structural representation of the target protein by using Preset option.

Tools Option

Different functions can be performed by tools option on the target protein for visualization and also for analyses.

Sequence

The sequence of the target protein can be seen by choosing the option **"Tool"** in the menu bar and **"sequence"** in the sub menu (Fig. **77**).

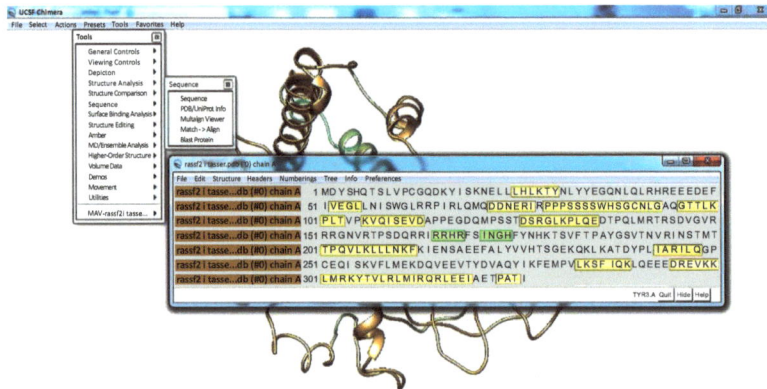

Fig. (77). Sequence of RASSF2.

Light Adjustment

For a better view of the target protein or insights of the protein, the user can change the intensity of the light. Select the **"Lighting"** tab in the **"Viewing"** window and the light sources and parameters will be displayed. The key light is the dominant brighter source of light and the fill light shown as a secondary source. The arrows in the right view allow to manipulating the lighting directions by using the mouse (Fig. **78**).

Interacting Residues Analyses

Firstly, open the protein file in Chimera, click on the **"SELECT"** option and select the ligand or the interacting partner. Click on the *TOOLS* option and select the **"STRUCTURE ANALYSIS"**. From the drop-down list of Structure Analysis, select the **"find Clashes/ Contacts"**.

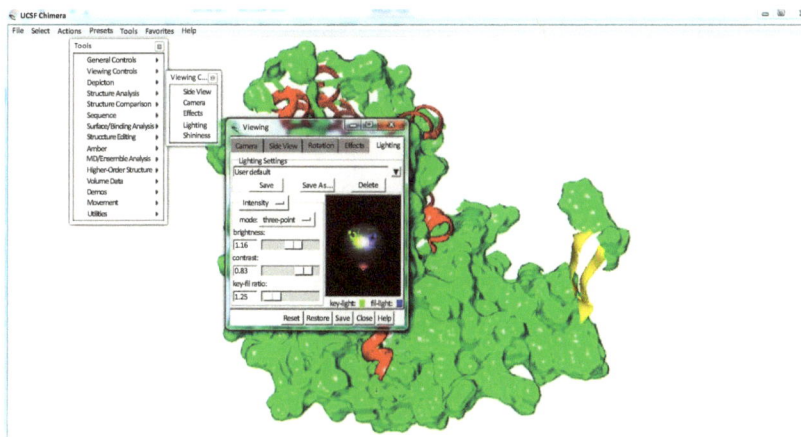

Fig. (78). Changing the intensity of light.

A new screen will appear and select the **"DESIGNATE"** option followed by **"OK"**. After designation, the interacting residues will be highlighted (Fig. **79**).

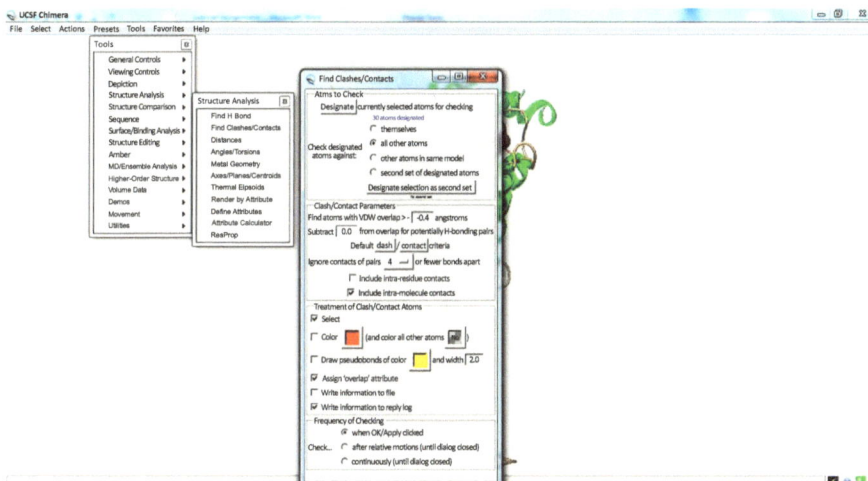

Fig. (79). Finding residues of RASSF2.

Select the **"ACTION"** option from the bar and select the **"LABEL"** from the dropdown list. Select the *RESIDUE* option and select **"1-LETTER CODE+SPECIFIER"**. There is another option of three letter code for the amino acid and the user can also use that according to the need for visualization.

Energy Minimization

Energy minimization is the significant process to minimize the protein for further drug design studies. To minimize the protein, open the file of the target protein in Chimera, go to the **"TOOLS"** option, and select the **"STRUCTURE EDITING"**, from the drop-down list select **"MINIMIZE STRUCTURE"**.

The minimization process window will appear on the screen and the user can change the **"STEEPEST DESCENT STEPS"** and **"CONJUGATE GRADIENT STEPS"** according to requirement. Here, we used 1000 steepest gradients and 1000 conjugate gradients for the minimization of RASSf2. Click on **"MINIMIZE"** button for minimization (Fig. **80**).

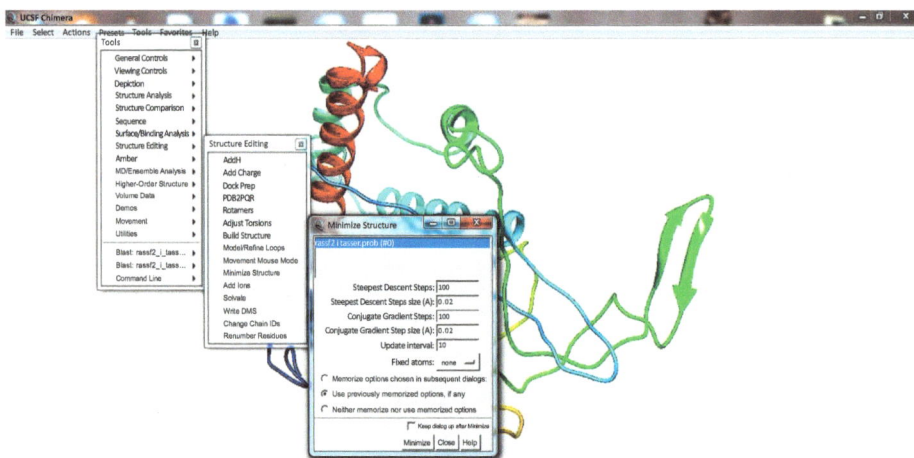

Fig. (80). Energy minimization of RASSF2.

LIMITATIONS OF STRUCTURE VISUALIZATION TOOLS

UCSF Chimera

• It takes much time to save Pov-Ray images.
• Sometimes, it goes to the non-responding state, often working on windows based operating system.
• There is no undo option available, to reverse any mistake.
• Protein-protein docking complexes are difficult to analyze and determine their interactions in Chimera.

It has various features for high quality graphics and structure editing, visualization, protein modeling and much more. Tools features, description, weblink and reference are mentioned in Table **6**.

Table 6. Visualization of predicted structure

Tool Name	Availability	Description	Work Flow	URL	Reference
UCSF Chimera 1.8	Freely available Tool	It is used for interactive visualization and analysis of molecular structures and related data.	+Open Model =Analyses =Visualizations	http://www.cgl.ucsf.edu/chimera/	Pettersen *et al.,* 2004

REFERENCES

Goddard, T.D., Huang, C.C., Ferrin, T.E. (2007). Visualizing density maps with UCSF Chimera. *J. Struct. Biol., 157*(1), 281-287.
[http://dx.doi.org/10.1016/j.jsb.2006.06.010] [PMID: 16963278]

Pettersen, E.F., Goddard, T.D., Huang, C.C., Couch, G.S., Greenblatt, D.M., Meng, E.C., Ferrin, T.E. (2004). UCSF Chimera--a visualization system for exploratory research and analysis. *J. Comput. Chem., 25*(13), 1605-1612.
[http://dx.doi.org/10.1002/jcc.20084] [PMID: 15264254]

Yang, Z., Lasker, K., Schneidman-Duhovny, D., Webb, B., Huang, C.C., Pettersen, E.F., Goddard, T.D., Meng, E.C., Sali, A., Ferrin, T.E. (2012). UCSF Chimera, MODELLER, and IMP: an integrated modeling system. *J. Struct. Biol., 179*(3), 269-278.
[http://dx.doi.org/10.1016/j.jsb.2011.09.006] [PMID: 21963794]

## CHAPTER 7

Molecular Docking Studies

Abstract: The molecular docking is a significant approach to computer-aided drug design coupled with structural biology. The ligand-protein docking analyses perform to predict the binding domains of a ligand with a protein.

Keywords: Autodock tools, Autodock vina, GOLD, Molecular docking, Protein ligand docking.

MOTIVATION

Molecular docking is an important technique to find out possible interactions between a protein and its counterpart *i.e.* ligand or protein. In this chapter, only protein - ligand interactions would be explored. The molecular docking normally uses the energy functions to analyze the fitting of a specific ligand into the binding pocket of the protein. If the binding pocket is not defined, it helps to identify those specific portions of the protein where a ligand can bind with high affinity. It has been observed that the molecular docking experiments help to decrease the search space for finding suitable ligands for a protein of interest. Here, two most widely used tools will be discussed – AutoDock and GOLD – both of them has a high success rate in identifying the ligands and the ligand binding pockets of a protein. AutoDock is freely available and can perform docking type experiments while GOLD is a tool package and requires the annual licensing fee while providing many advanced features.

In this Chapter, we will follow the methodology of section 07 mentioned in Fig. (**3**).

MOLECULAR DOCKING ANALYSES

The aim of molecular docking analyses is to identify the binding pattern and the relative binding specificities. AutoDock Vina docking software can be used to investigate how ligand binds to the target protein, the binding conformation, functionally interacting residues and best structural information.

Sheikh Arslan Sehgal, Rana Adnan Tahir, A. Hammad Mirza & Asif Mir

AUTODOCK

AutoDock is an automated blind docking tool and can also be used for targeted docking studies. It is designed to predict how small molecules including substrates or drug candidates, bind to a receptor having 3D structure. AutoDock is not intended to perform protein-protein docking; it can only perform ligand (small molecule) and receptor (protein) dockings studies. AutoDock implements the genetic algorithm and it does not provide better results for flexible docking. So, AutoDock can better perform for rigid docking experiments. Current distributions of the AutoDock consist of two generations of software: AutoDock 4 and AutoDock Vina (Goodsell *et al.,* 1996).

AutoDock Applications

AutoDock has applications in:

- X-ray crystallography
- Structure-based drug design
- Lead optimization
- Virtual screening (VS)
- Combinatorial library design
- Chemical mechanism studies

AutoDock 4

AutoDock 4 actually consists of two main programs: i) AutoDock which performs the docking of a ligand to a set of grids describing the target protein and ii) AutoGrid that pre-calculates the grids upon which the ligand would try to fit into (Chang *et al.,* 2010).

Requirements

Input

The input files required for AutoDock is the .pdb file of the target protein and .pdbqt file of ligand molecule. Here, we will use the predicted structure of RASSF2 as the target protein and we choose a ligand from the literature for RASSF2. In this chapter, we tried to write the commands to follow for the smooth run of analyses and also try to make it more convenient for the beginners. We performed all the mandatory steps for the identification of a potential ligand for the target protein and also keep this in mind to make it easy for the beginners.

Preparing the Protein

To upload the .pdb file of RASSF2, we use the option of File – Read Molecule and upload the structure of RASSF2 (Fig. **81**).

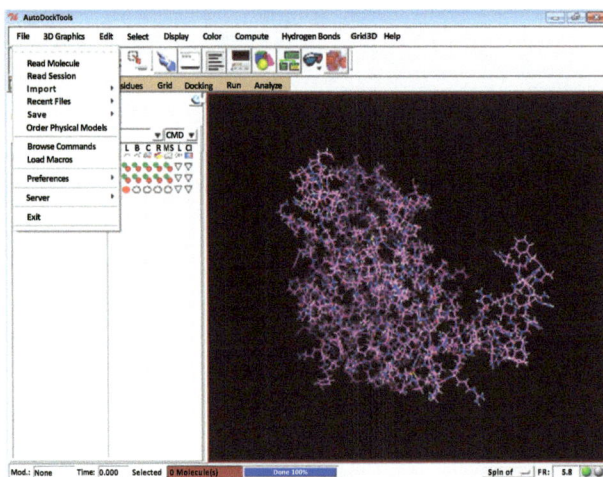

Fig. (81). Preparing the protein.

Add Hydrogen's

Edit → Hydrogens→ Add → choose "All hydrogen's" (Fig. **82**). Usually, the protein structure do not have hydrogen atoms and it is significant for reliable results to add the hydrogen atoms in the protein molecule.

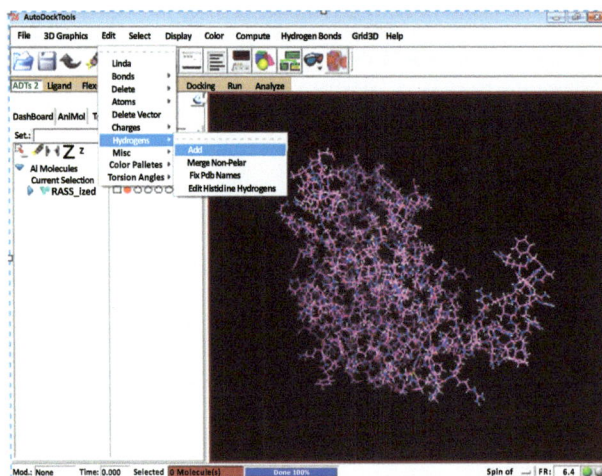

Fig. (82). Adding hydrogen.

Hide Protein

Now, the protein or receptor is prepared. Right-click on the Molecules at the dashboard and click to hide the protein (Fig. **83**). The purpose of hiding the protein is to make the screen clear for the ligand preparation to avoid the mistakes. The user can skip this step but it's better to perform.

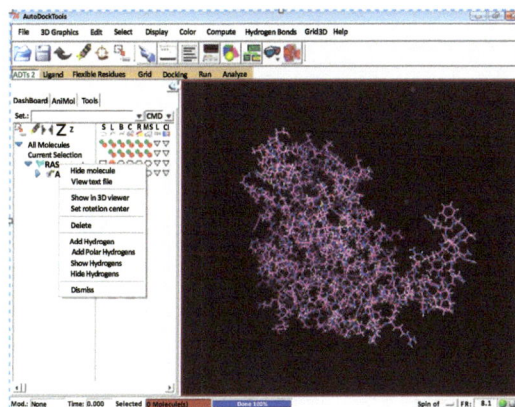

Fig. (83). Hiding the protein.

Preparing the Ligand

Opening File

Ligand → Input → Open → All Files → [choose file] → Open (Fig. **84**). Usually, the FDA approved drug molecule, screened libraries to identify a molecule, novel designed molecule or molecule reported in the literature are used as a ligand depends on the nature of the experiment.

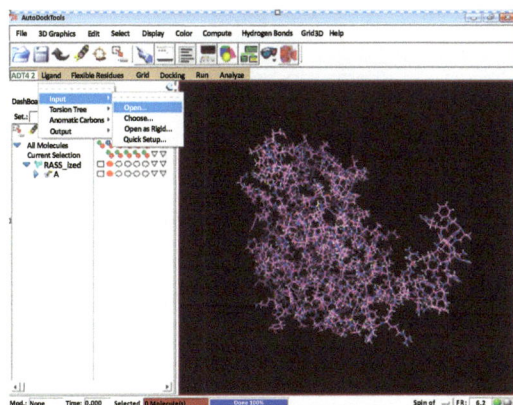

Fig. (84). Opening the file.

Define Torsions

Ligand → Torsion Tree → Set Number of Torsions to choose the number of rotatable bonds that move the 'fewest' or 'most' atoms (Fig. **85**). The defining of the torsions also depends on the nature of the experiment.

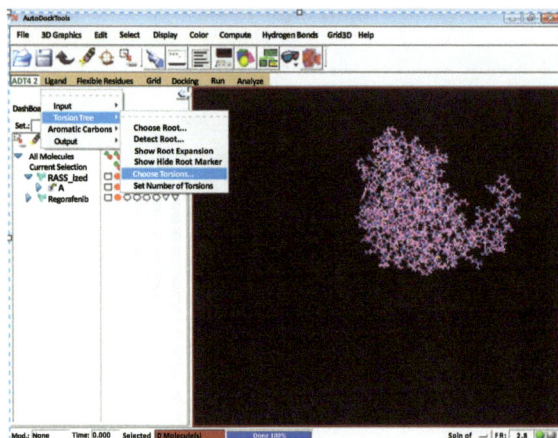

Fig. (85). Defining the Torsions.

Choose the number of rotatable bonds that move the 'fewest' or 'most' atoms. It is better to make all rotatable bonds to zero for rigid docking. For flexible docking, the user must know about the rotatable bonds that should rotate and just let those bonds rotatable and put all other bonds non-rotatable. Generally, AutoDock is considered better for the rigid docking, so it is better to make all the rotatable bonds to non-rotatable (Fig. **86**).

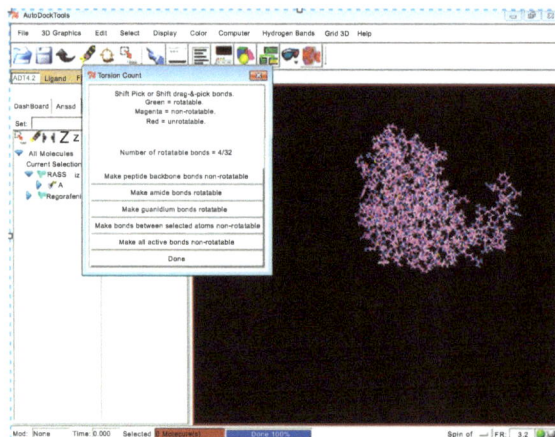

Fig. (86). 'Fewest' and 'most' atoms.

Save Ligand File

Ligand → Output → Save as PDBQT → save with Ligand.pdbqt (Fig. **87**). The pdbqt file has the gasteiger charges as per the requirement of the AutoDock.

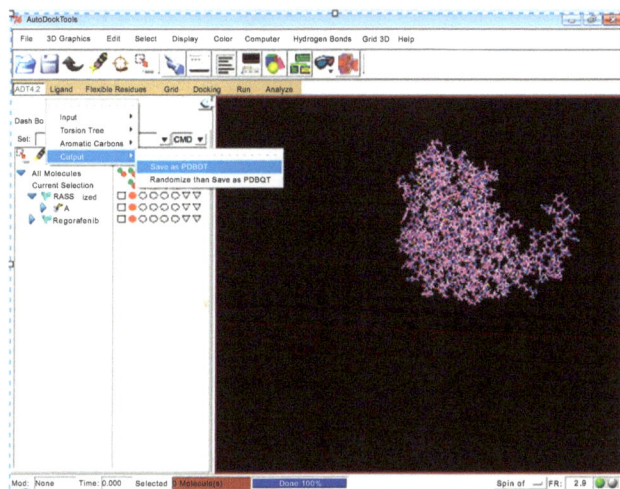

Fig. (87). Saving the Ligand File.

Hide the Ligand

The pdbqt file of the ligand molecule was prepared and by using the right-click on the "Molecule hide" option to hide the ligand molecule (Fig. **88**). This step is also optional.

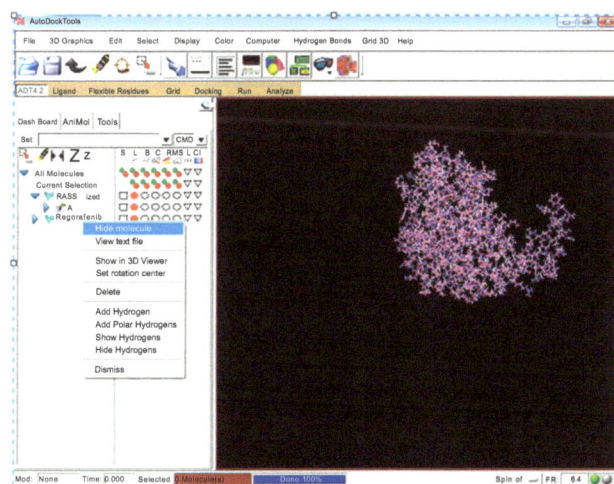

Fig. (88). Hiding the Ligand.

Running AutoGrid Calculation

Open the Protein

Grid → Macromolecule → Open → choose receptor .pdb file with add hydrogen (Fig. **89**). Now, the receptor molecule selected for the docking studies.

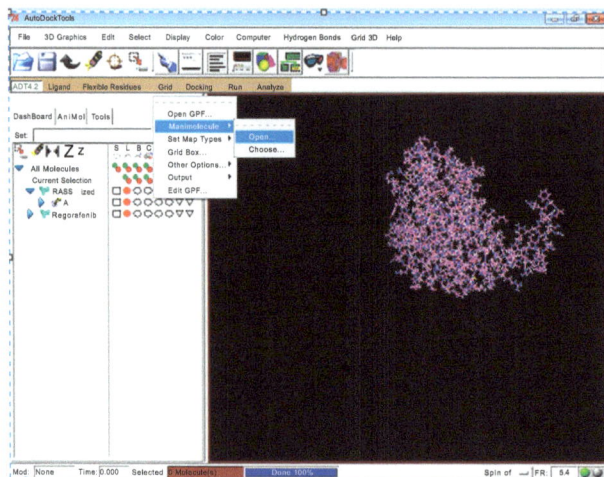

Fig. (89). Opening protein for AutoDock calculations.

Prepare Grid Parameter File

Grid →Set map type →choose ligand (choose the ligand already opened) (Fig. **90**). Ligand also has to choose for analyses.

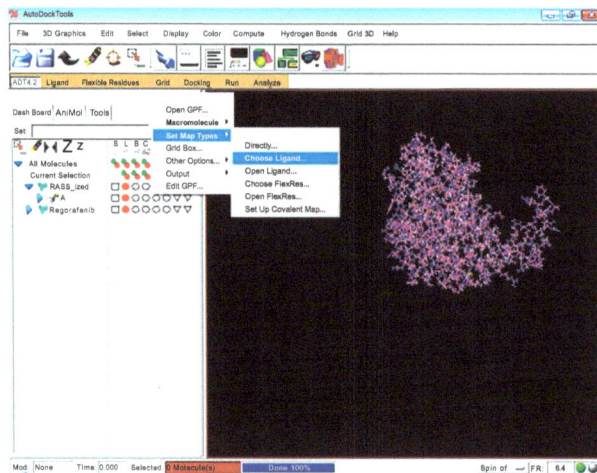

Fig. (90). Preparation of Grid parameter file.

Set Grid Properties

Grid → Grid Box → [Set the grid dimensions, spacing, and center] → File → Close Saving Current (Fig. **91**). Usually, for blind docking studies, the grid covers the whole receptor molecule to find out the specific position for the ligand interaction with high affinity. For targeted docking, the grid only covers the specific area where the user needs to dock the ligand molecule. We used blind docking studies here and choose the complete receptor for docking studies.

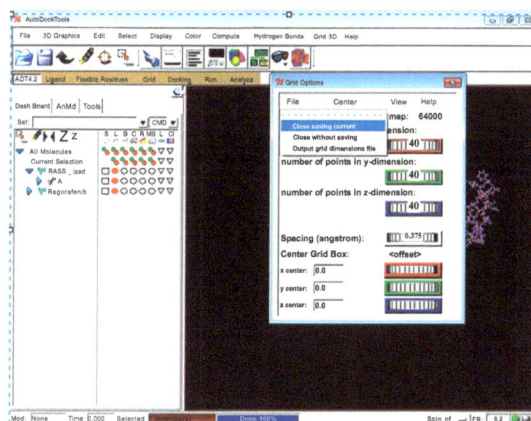

Fig. (91). Setting grid properties.

Save the Grid Settings

Grid → Output → Save GPF (grid parameter file) → save as grid.gpf (Fig. **92**). The file format for the grid file should be .gpf.

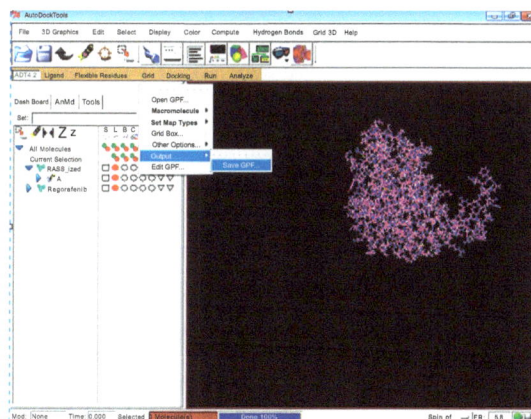

Fig. (92). Saving grid settings.

Preparing the Docking Parameter File (.dpf)

Specifying the Molecule

Docking → Macromolecule → Set Rigid File name → [choose grid.pdbqt] (Fig. 93). The grid file is chosen to use the specified values mentioned in the grid file.

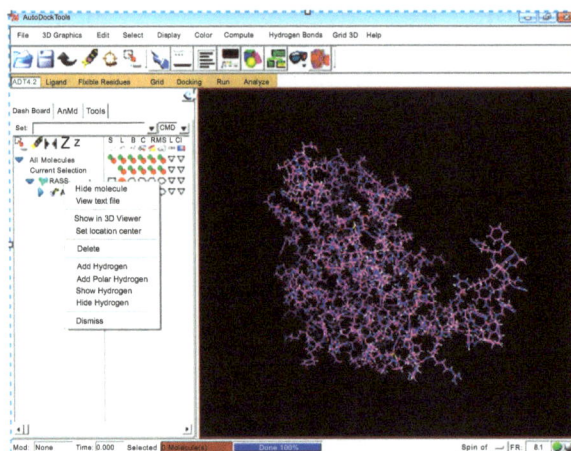

Fig. (93). Preparing .dpf File.

Specifying the Ligand

Docking → Ligand → Choose → [choose Ligand.pdbqt] → select ligand → parameter window open → by default parameter → Accept (Fig. **94**). We used the default parameter for the ligand here but the user can change as per requirement.

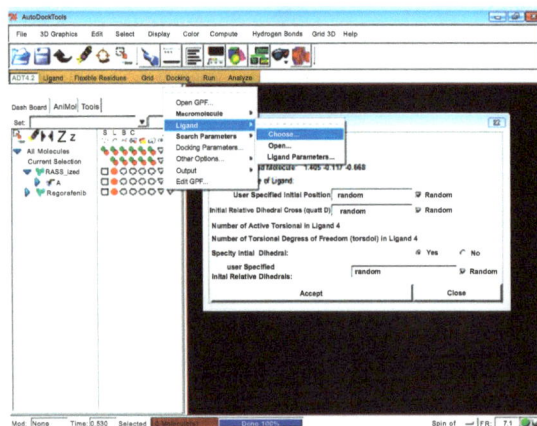

Fig. (94). Specifying the Ligand.

Choosing Docking Method

Docking → Search Parameters → Genetic Algorithm → number of genetic runs→ 50 runs → Accept (Fig. **95**). The selection of the algorithm is a significant point in docking studies and the Genetic algorithm is widely used for the molecular docking studies. The run of the analyses can be changed and 100 runs are considered reliable for the docking studies. The remaining parameters can also be changed as per the nature of the experiment.

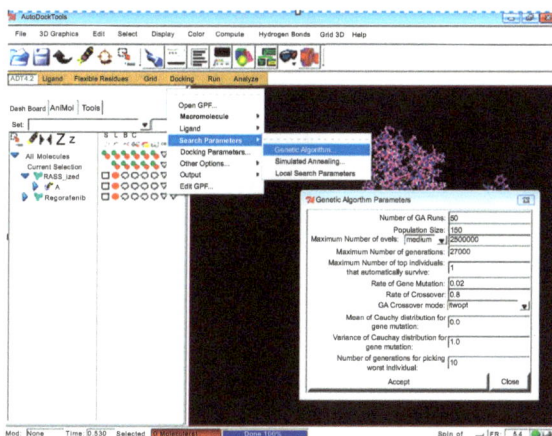

Fig. (95). Choosing Docking Method.

Setting Docking Parameters

Docking → Docking Parameters → choose the defaults (Fig. **96**). We used the default docking parameters for this analysis.

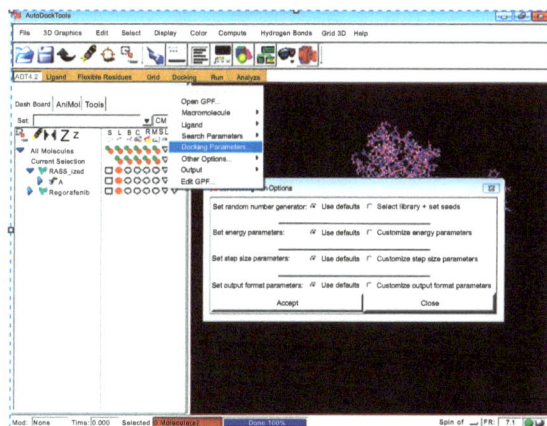

Fig. (96). Setting docking parameters.

Docking → Output → Lamarckian GA → save as results.dpf (Fig. **97**).

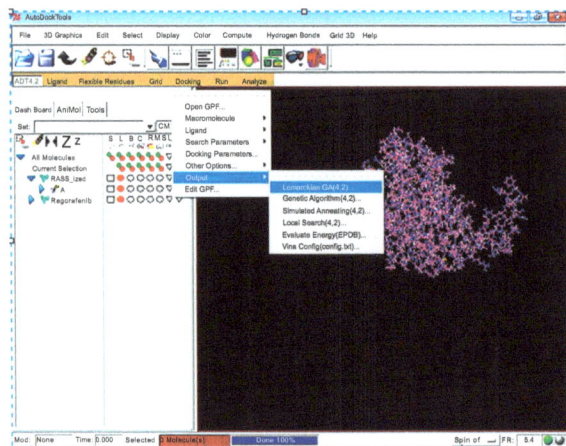

Fig. (97). Docking Parameters.

Running AutoDock4

Running

Run → Run Autogrid→ Launch

Run → Run AutoDock → Launch (Fig. **98**). For the successful running of the docking analyses, both the files should launch and run. It will take time depends on the chosen number of runs and also the nature of the experiment.

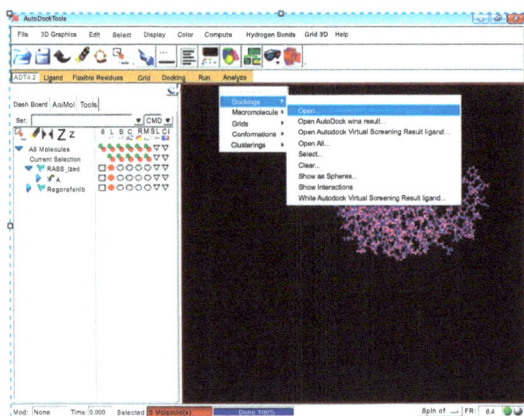

Fig. (98). Running AutoDock 4.

Docking Results

The analyses of docking results also perform by AutoDock. The results will be in

.dlg file and .dlg file will be uploaded to AutoDock for analyses.

Analyze → Dockings → Open → choose results.dlg

Analyze →Macromolecule→ Open → choose .pdbqt file (Fig. **99**). The automatically generated pdbqt file of the receptor will be used here for analyses.

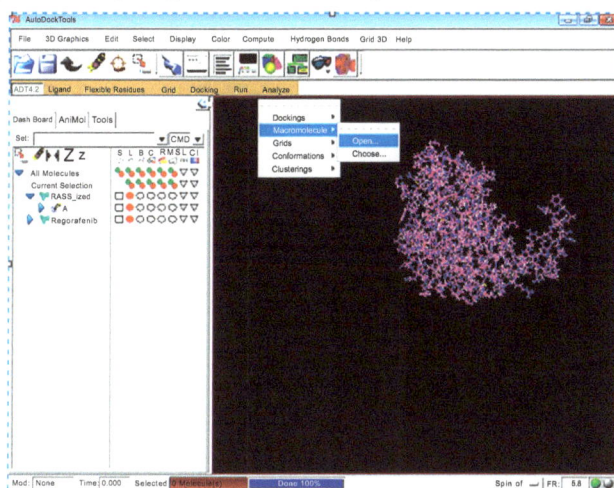

Fig. (99). Choosing file for analyses.

Analyze → Conformations → Play, ranked by energy → click on & sign → tick the "Show info" → write complex (Fig. **100**). All the 50 complexes will be played by energy ranking for the analyses.

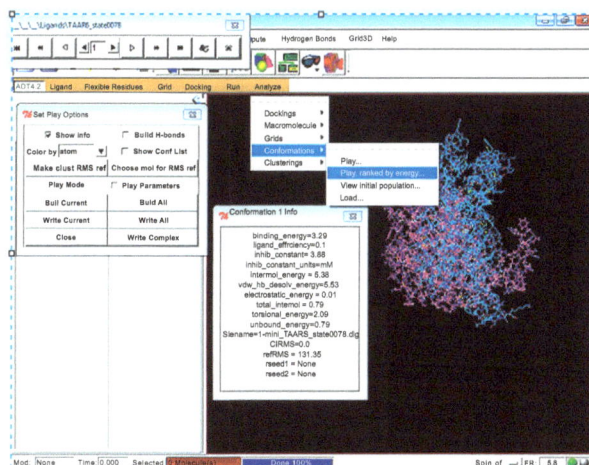

Fig. (100). View of the Complex.

The reliable complex has to save by using the Write Complex option in pdb format to visualize it in Chimera or any other visualizing tool. The visualization and interaction analyses by Chimera has been mentioned and described in the previous chapter.

AUTODOCK VINA

Introduction

AutoDock Vina is an automated procedure for predicting the interaction of the ligands with the bio-macromolecular targets (Morris *et al.*, 2009).

Getting Started With AutoDock Vina

First of all, prepare the protein and ligand by using AutoDock as described before. Click on AutoDock Icon for docking analyses by using the AutoDock Vina (Fig. **101**).

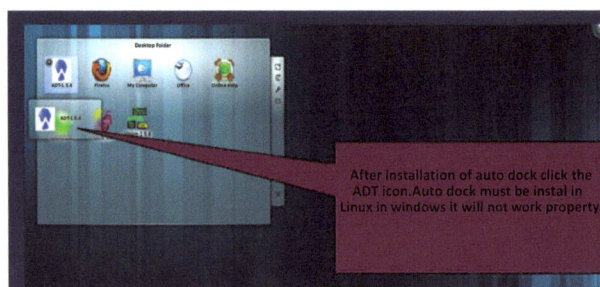

Fig. (101). Starting AutoDock.

The interfaces will appear for Linux (Fig. **102**) and for windows (Fig. **103**). You can register the AutoDock with the provider or you can also skip this registration.

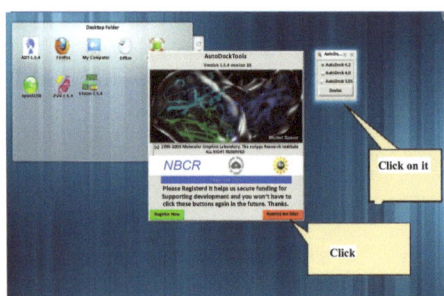

Fig. (102). AutoDock Interface on Linux.

Fig. (103). Opening of AutoDock.

The interface of the AutoDock will be appeared to start the docking analyses (Fig. **104**).

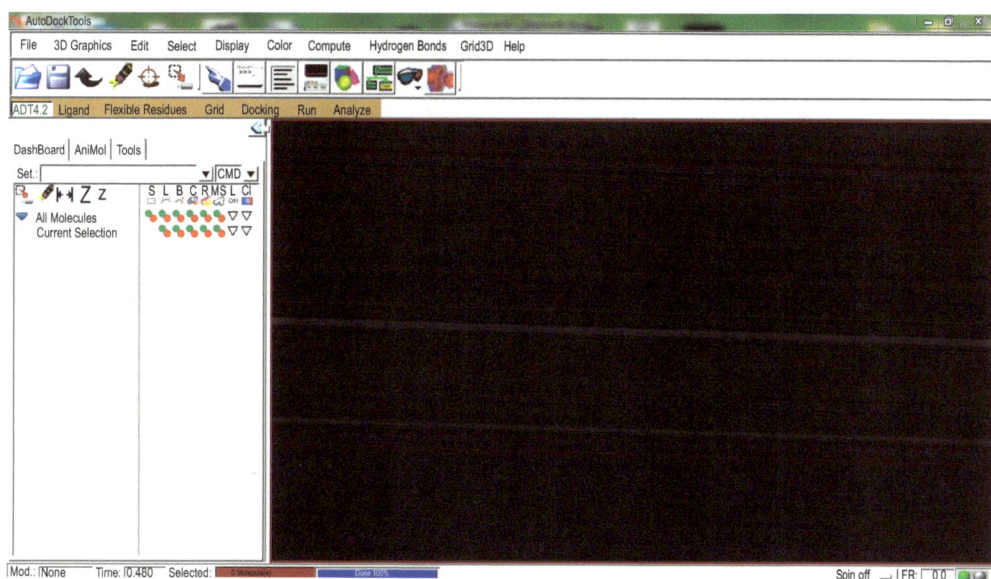

Fig. (104). AutoDock window Interface.

Protein Preparation

The predicted structure of the RASSF2 was utilized and uploaded by using the File option followed by the Read Molecule --> RASSF2 minimized.pdb (Fig. **105**).

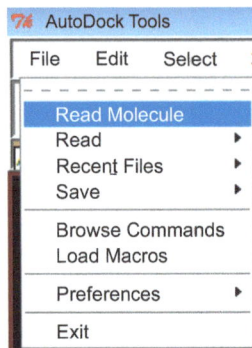

Fig. (105). Opening the Minimized RASSF2 PDB file.

The protein RASSF2 uploaded on the AutoDock for analyses and to prepare the protein (Fig. **106**).

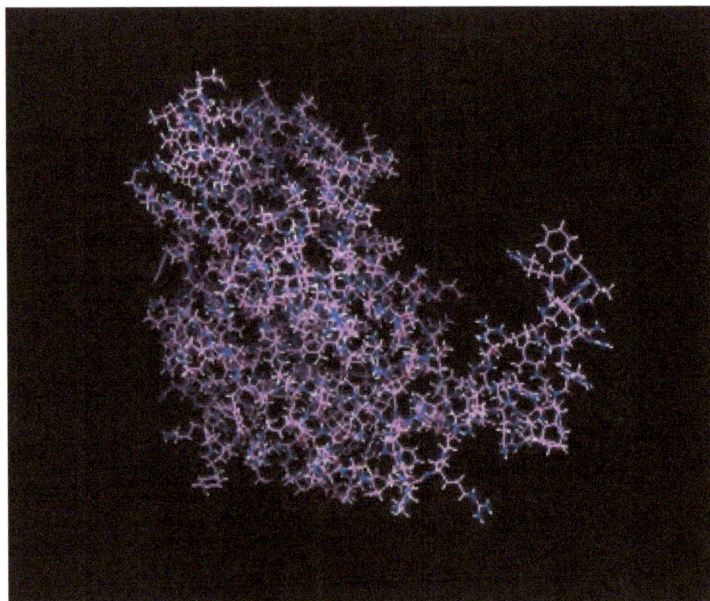

Fig. (106). View of RASSF2 in AutoDock Vina Interface.

The RASSF2 was uploaded and colored by c Color option. From the given options, click on By Atom type and a new screen will appear (Fig. **107**).

The user can choose any option of choice but here we selected 'All Geometries' and click on 'OK' (Fig. **108**).

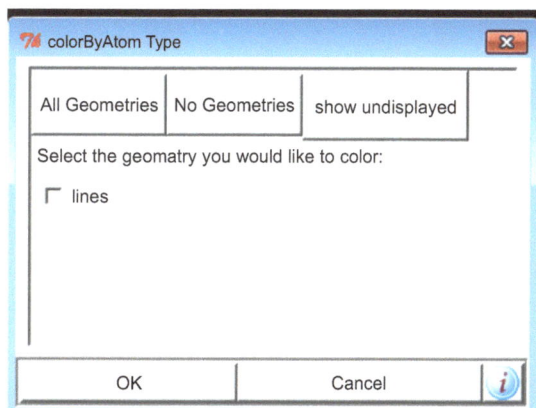

Fig. (107). Coloring the Protein (RASSF2).

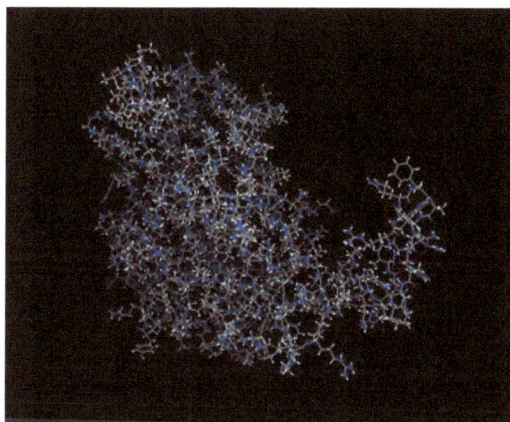

Fig. (108). All atoms of RASSF2 colored by type.

Sometimes, the water molecules are present in the protein and it is good to remove the water molecules. The water molecules from RASSF2 were also removed by clicking on the option from the menu bar.

Select → Select from string → [write HOH* in "Residue"

Line and * in the "Atom" line] → Add → Dismiss → Edit → Delete → Delete Atom Set.

Add the hydrogen atoms to RASSF2 for docking studies by

Edit → Hydrogen → Add → Polar only → noBondOrder → Yes → OK (Fig. **109**)

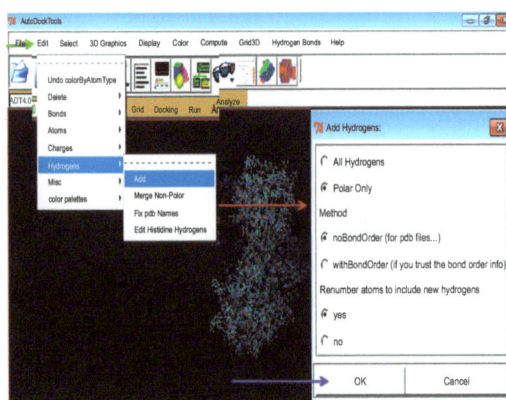

Fig. (109). Adding Hydrogen to RASSF2.

The hydrogen atoms will be added and after that select the option Polar Only and we used the remaining parameters as default.

Usually, by experimental structures, some or few atoms usually miss from the structure and to overcome this issue we added the missing atoms.

Edit → Misc. → Check for missing atoms

Edit →Misc. → Repair missing atoms (Fig. **110**).

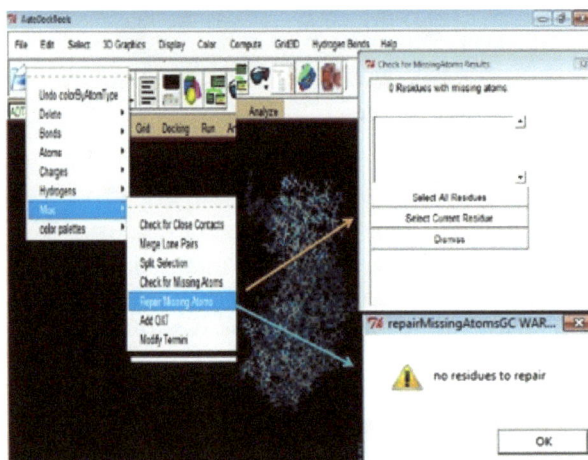

Fig. (110). Repairing the missing Atoms.

There is no need to repair the missing atoms if there are no missing atoms. Repair only, if some residues will be missing otherwise only perform the step to check the missing residues. To repair the receptor or adding the missing atoms, the receptor was saved for further analyses by utilizing *File - save - Write PDB* options (Fig. **111**).

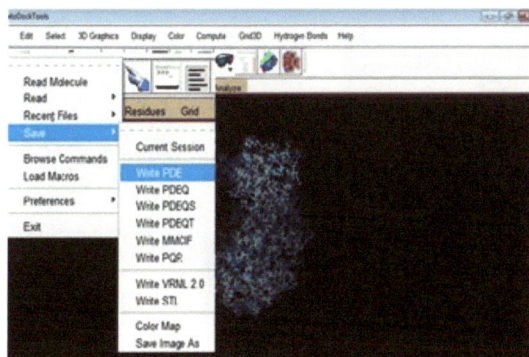

Fig. (111). Saving the Protein.

Ligand Preparation

To prepare the ligand file for the docking experiment, we opened the ligand by

Ligand → Input → Open → All Files → [choose file] → Open (Fig. **112**)

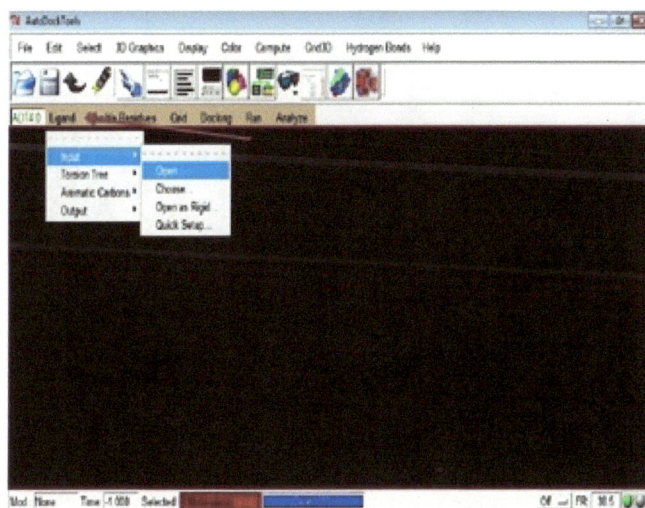

Fig. (112). Preparing the Ligand.

AutoDock automatically computed the Gastegier charges, merges non-polar hydrogen atoms and assigns Autodock Type to each atom (Fig. **113**).

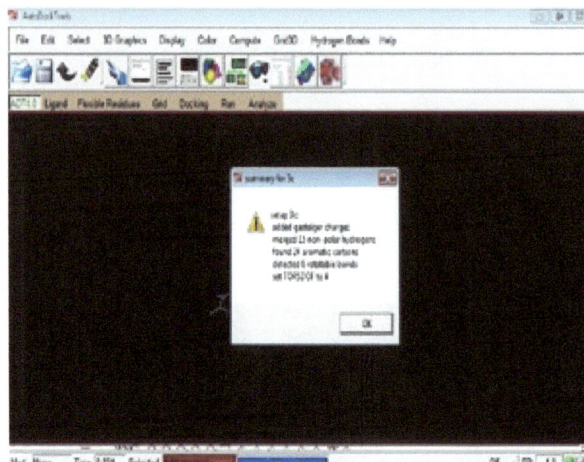

Fig. (113). Giving Ok Command Prompt.

The torsion angles have to describe or set according to the nature of the experiment and we used the detect root option

Ligand → Torsion Tree → Detect Root (Fig. **114**)

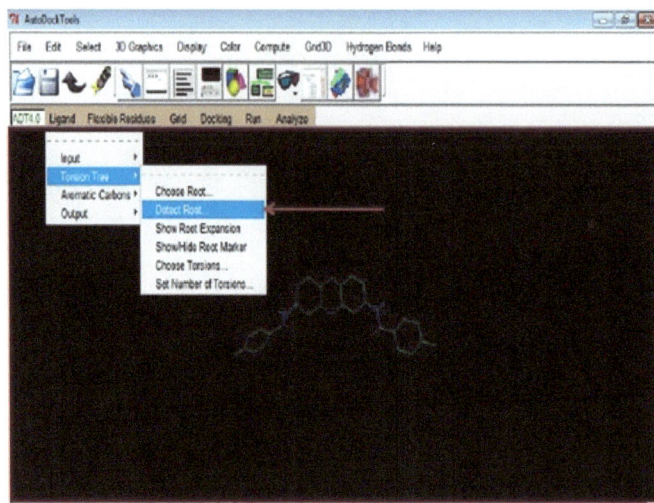

Fig. (114). Detecting the Root.

Ligand → Torsion Tree → Choose Torsions (Fig. **115**)

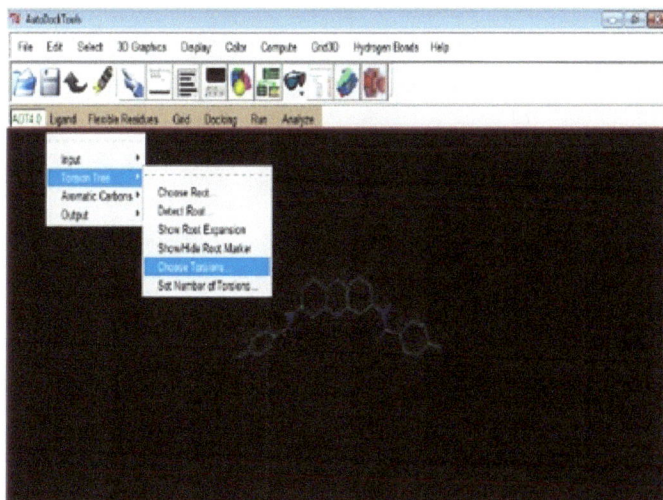

Fig. (115). Defining the Torsions.

Click on choose Torsions and a new window will appear with different options (Fig. **116**).

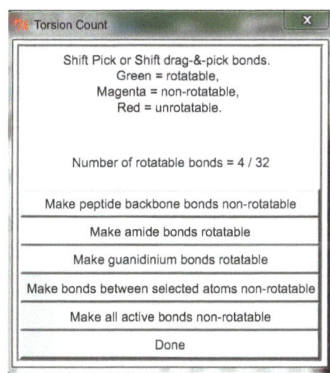

Fig. (116). Torsion Count.

As discussed earlier in this chapter regarding the rotatable bonds, we used rotatable bonds=0 by clicking on the first option (Fig. **116**) and click the last option (make all rotatable bonds non-rotatable) (Fig. **116**).

Ligand → Torsion Tree → Set Number of Torsions → check the option named "most" (Fig. **117**) to set the torsion angles of the ligand.

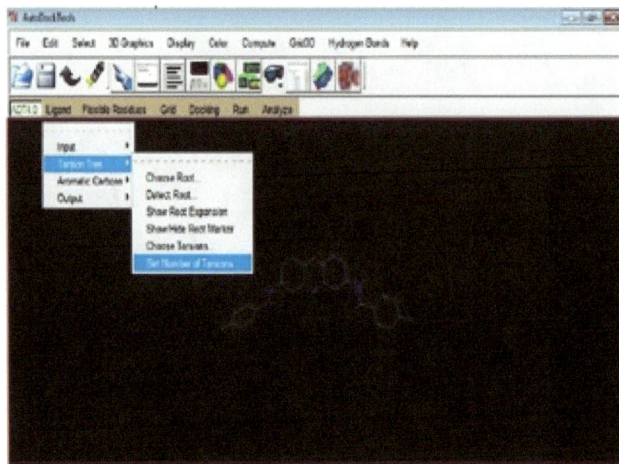

Fig. (117). Setting number of Torsions.

Generate the pdbqt file of ligand by **Ligand → Output → Save as PDBQT** (Fig. **118**) having charges.

Preparation of PDBqt of Protein

The pdbqt file of the protein was also generated for the analyses by

Grid → Macromolecule → Open → [open RASSF2.pdb] (Fig. **119**)

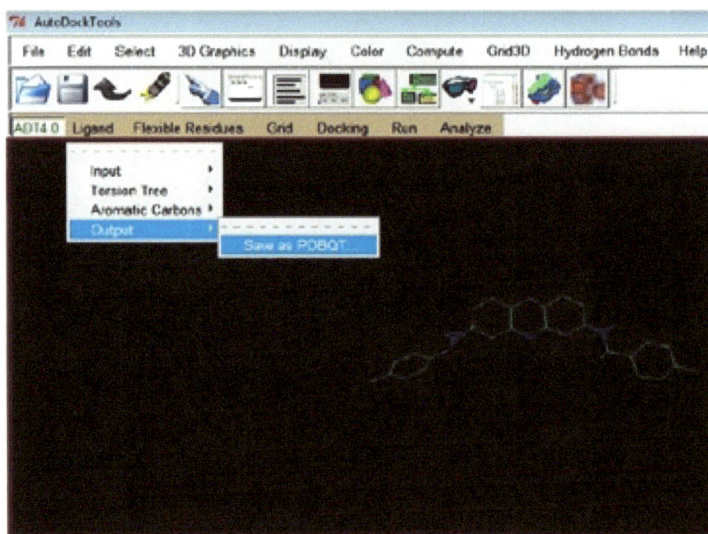

Fig. (118). Saving the Ligand File.

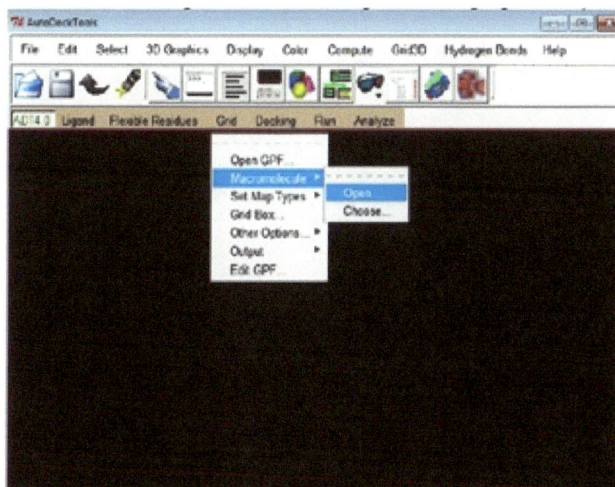

Fig. (119). Open RASSF2 PDB file.

Choose pdb of the RASSF2 which was prepared and discussed above in this chapter

AutoDock will automatically add the charges and also merge the hydrogen atoms (Fig. **120**) to receptor.

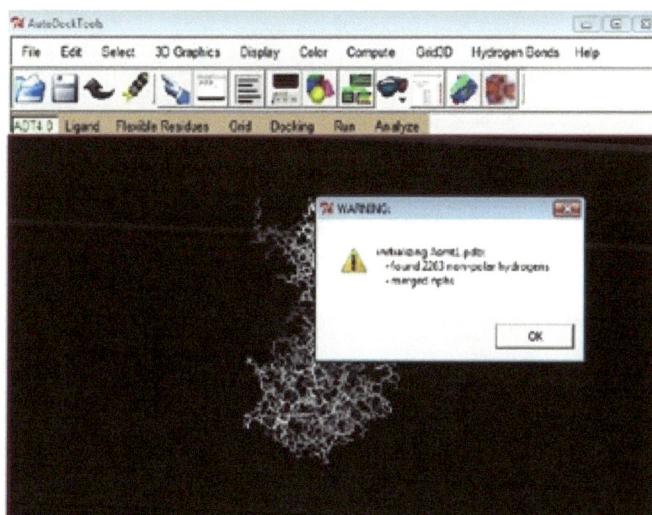

Fig. (120). Adding charges and Merging Hydrogen.

The prepared receptor having chargers have to save in pdbqt as **protein .pdbqt.**

To set the grid box for docking studies by **Grid → Grid box** (Fig. **121**).

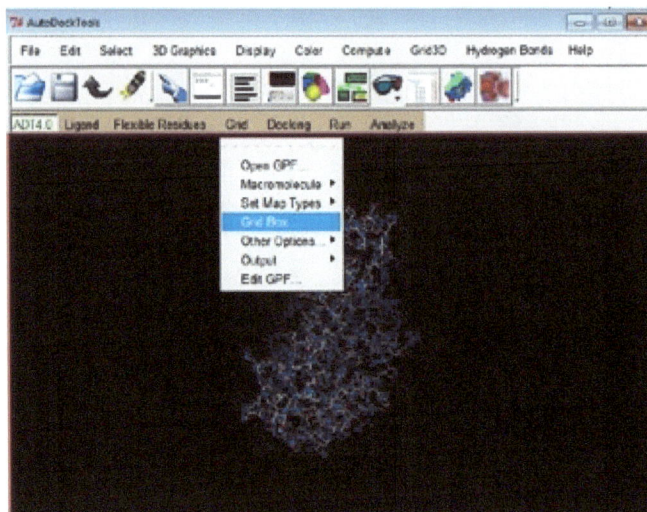

Fig. (121). Setting Grid Box.

The new window will appear having a number of points in x, y and z dimensions and the values for x center, y center and z center (Fig. **122**).

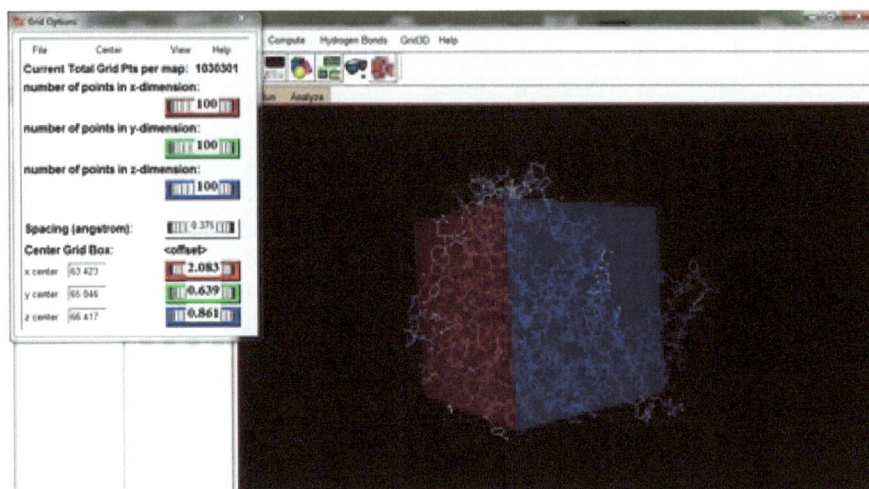

Fig. (122). Setting the Co-ordinates.

By default, the value of the number of points in x, y and z dimension will be 40 but for blind docking, we have to cover the complete protein and we used 100 as the dimension. It is better to note down the values of x center, y center and z center otherwise the user has to memorize the values for the configuration file. A configuration file has to write in .txt format by opening a word document or notepad and write down the name of the specified files as RASSF2.pdbqt (receptor), regorafenic.pdbqt (ligand), coordinates values (Fig. **123**) and save the file in.txt format.

Fig. (123). Conf.txt File.

AutoDock Vina for Linux

Before running the AutoDock Vina, for the files "RASSF2.pdbqt (receptor), regorafenib.pdbqt (ligand) and conf.txt" must be present in the directory folder of the AutoDock vina and the .exe of vina should also be there. After this, run the following command at the command prompt.

./vina.exe --config conf.txt --log log.txt (Fig. **124**)

Fig. (124). Command Prompt during Vina Docking.

AutoDock Vina for Windows

The appropriate files have to place in the respective directory and run the following command at the command prompt.

vina.exe --config conf.txt --log log.txt (Fig. **125**)

Fig. (125). Running AutoDock Vina on Windows.

The docking analyses will complete and the output file will be the out-file of .pdbqt having the 3D predicted models along with their binding affinities. To generate the .pdb file of the docked complex, copy the model having lowest binding energy values and paste at the end of the original .pdb file of the receptor (RASSF2) (Copy HETATOMS after the "root" and have to copy before the "endroot"). Paste the HETATOM at the end of the .pdb file after TER. Finally, the docked complex is ready for the visualization and analyses. The interacting residues can be analyzed by UCSF Chimera as discussed in the previous chapter.

GOLD

Introduction

GOLD reliably identifies the correct binding mode for a large range of test set cases and is highly configurable allowing you to take full advantage of your knowledge of a protein-ligand system in order to maximize the docking performance. GOLD has a wide range of available scoring functions and customizable the docking protocols. GOLD provides consistently high performance across a diverse range of the receptor types. GOLD accounts for receptor flexibility through the side-chain flexibility and most importantly ensemble docking.

Getting Started with GOLD

GOLD is an annually licensed software for the molecular docking studies and other computational studies. Here, we will focus on the molecular docking expertise of the GOLD and how to perform the molecular docking through GOLD. Install the GOLD software and click on HERMES icon (Fig. **126**) to start the GOLD.

Fig. (126). After clicking on Hermes icon, the Hermes Window will appear.

The "GOLD" interface (Fig. **127**) will appear and select the Hermes visualizer - "Wizard" from the resulting menu. The new interface will appear (Fig. **128**)

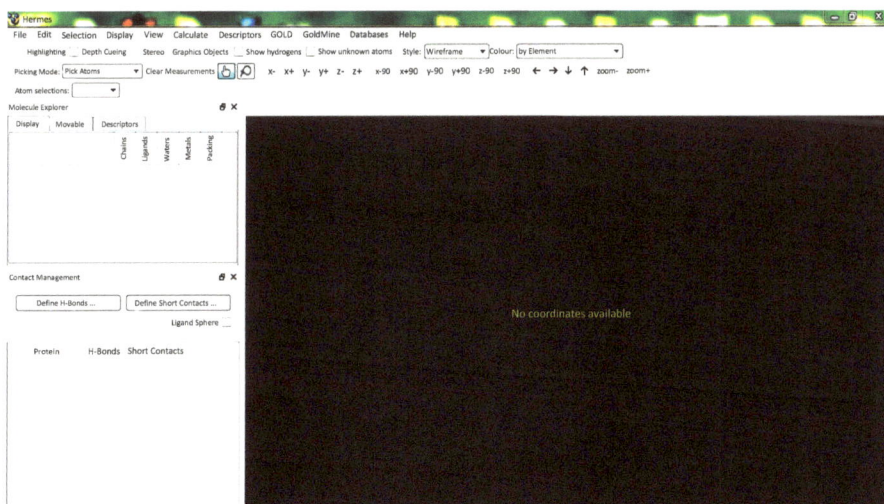

Fig. (127). GOLD Interface.

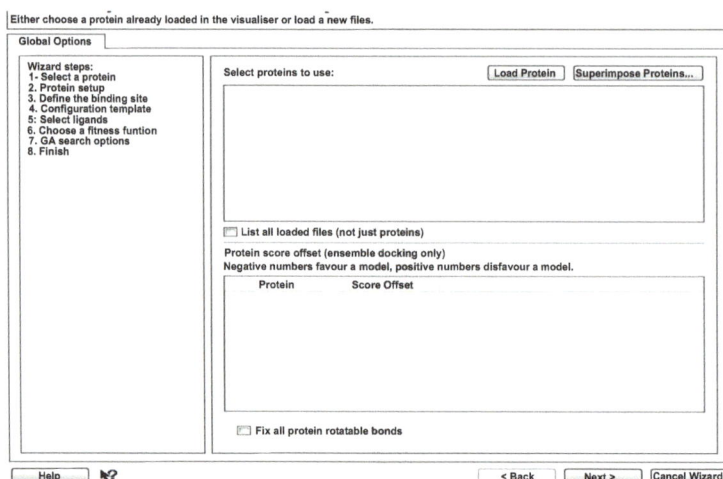

Fig. (128). Protein loading.

Click the "Load protein" option for the loading of the target protein (RASSF2) and as a result, the selected protein will be loaded and its name will appear (Fig. **129**).

Click on the Next option and the protein setup window will appear. There will be the option of add hydrogen and by selecting this option, the hydrogen atoms will be added. There will also an option to remove the water molecules or HETATOMs (LIGAND) present in the selected protein (Fig. **130**).

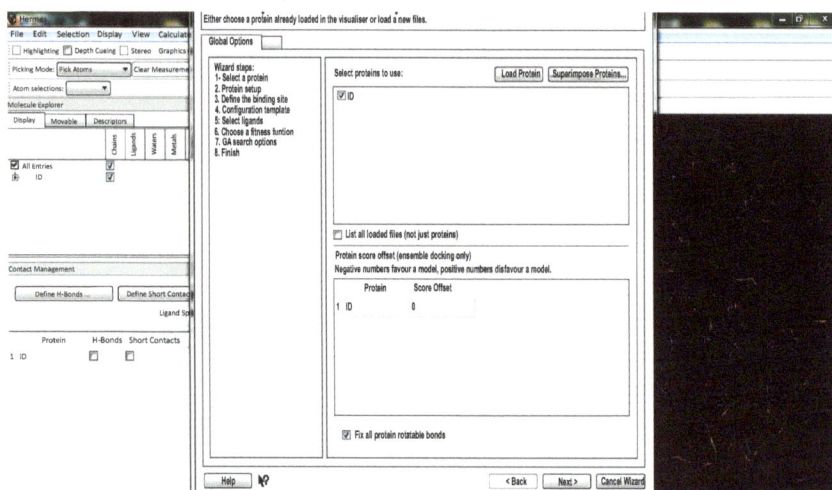

Fig. (129). Protein preparation.

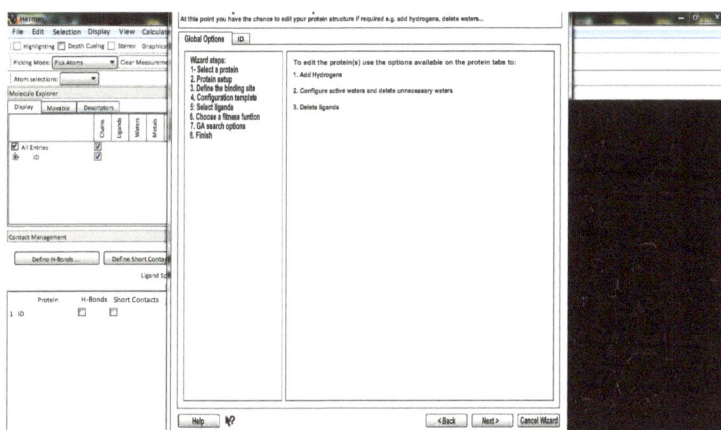

Fig. (130). Adding hydrogen atoms.

After this step, click the "Next". Now, the different options will appear to define the binding site of the selected protein (Fig. **131**).

For targeted docking, check the first option and minimize the wizard. On Hermes window, there will be selected protein option to click. Select the atom by clicking on that specific atom(s) for the targeted docking. The residue will become yellow after selecting and the selected label name will be written under the option of "Atom". Now while keeping the default value of 10 Å, click next. For blind docking, use the "select all atoms within" option and increase the value to cover the complete protein (Fig. **131**).

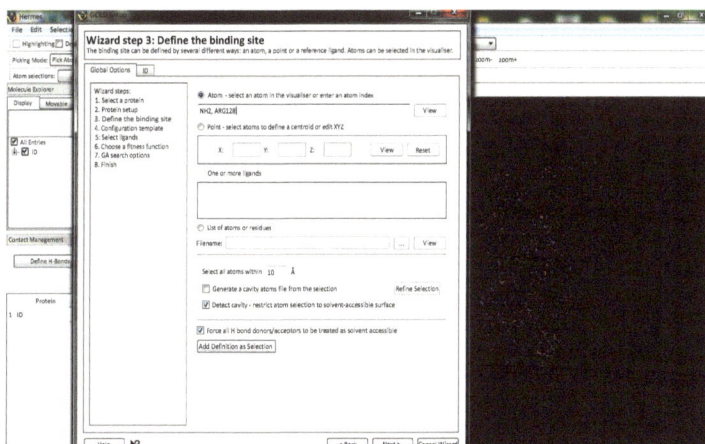

Fig. (131). Receptor binding sites.

The selection of the algorithm depends on the nature of the experiment and protein. Select the algorithm by "Configuration template" followed by "goldscore_p450_csd" and click on next (Fig. **132**).

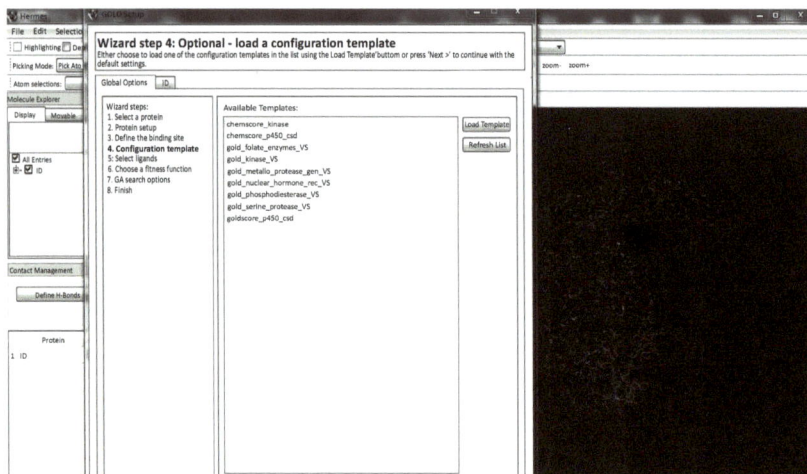

Fig. (132). Selection of algorithm.

After configuration the template, click on the "Select ligands" option and upload the scrutinized ligands in .mol2 file (Fig. **133**).

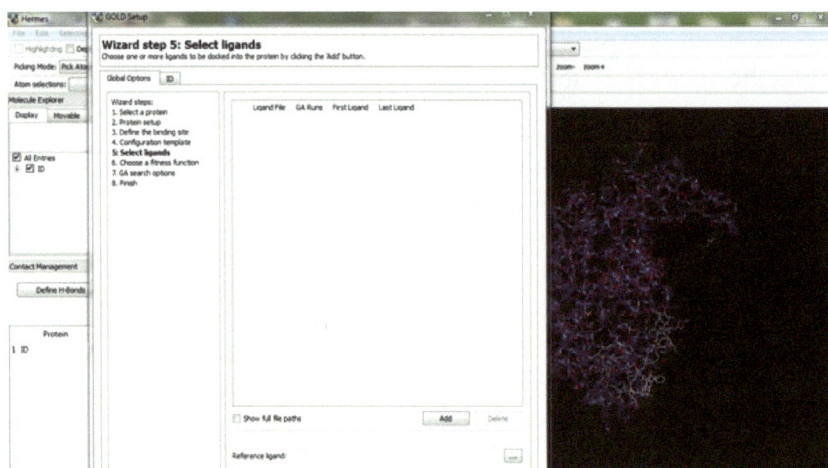

Fig. (133). Ligand selection.

The "Regoravenib" ligand was selected for the docking studies (Fig. **134**).

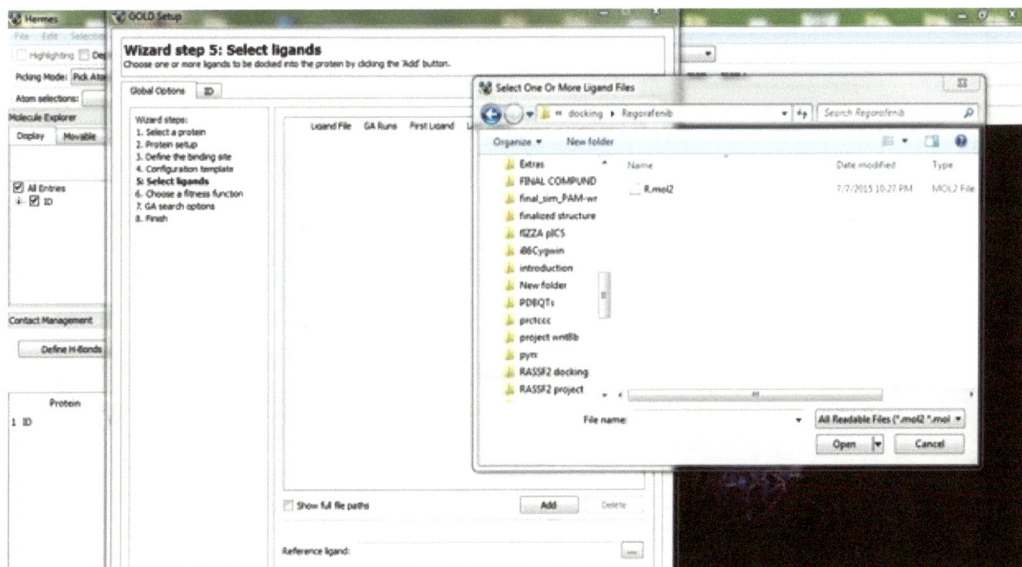

Fig. (134). Loading of ligand.

The ligand "R.mol2" has been loaded (Fig. **135**).

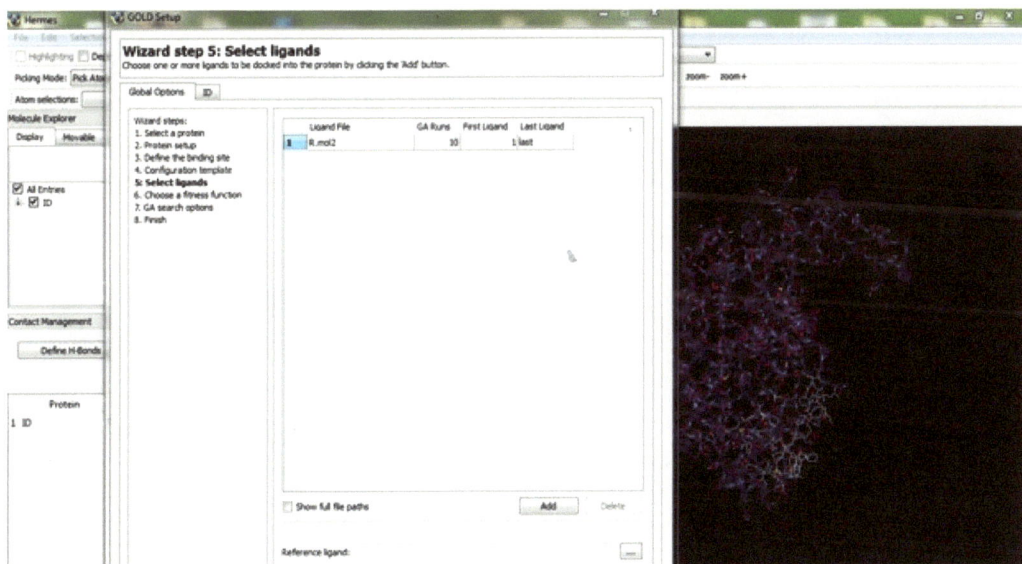

Fig. (135). Ligand loaded.

Click on the next, the new window "Choose a fitness function" will appear. We selected the "Goldscore" as the fitness function for this analysis (Fig. **136**).

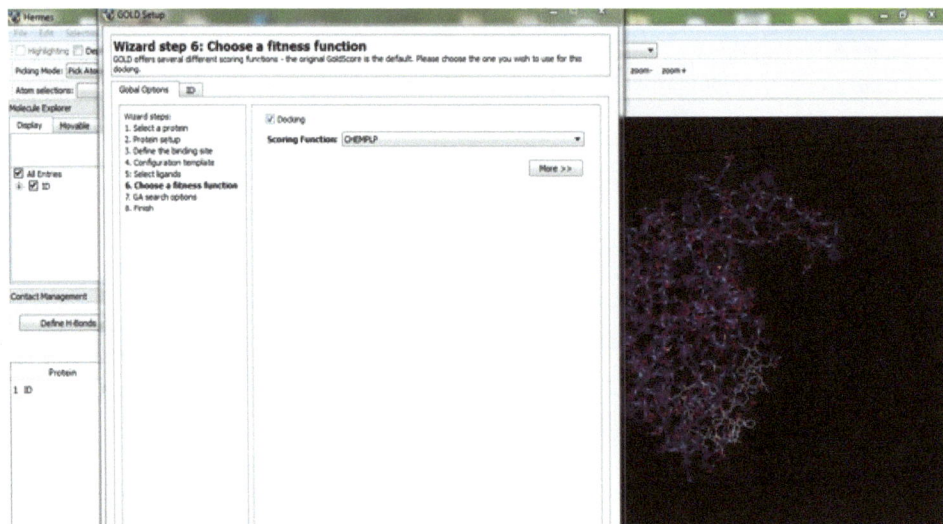

Fig. (136). Selection of fitness score.

After clicking on next, the new window will appear which showed following "GA search options". Select the "slow" option for most accurate results (Fig. **137**). Click on the "advanced" followed by "Global options" and the user will be able to check out all the selected parameters (Wizard, Templates, proteins, define binding sites, select ligands, configure waters, ligand flexibility, fitness and search options, GA settings, output options, Parallel GOLD, constraints and Atom typing) (Fig. **138**).

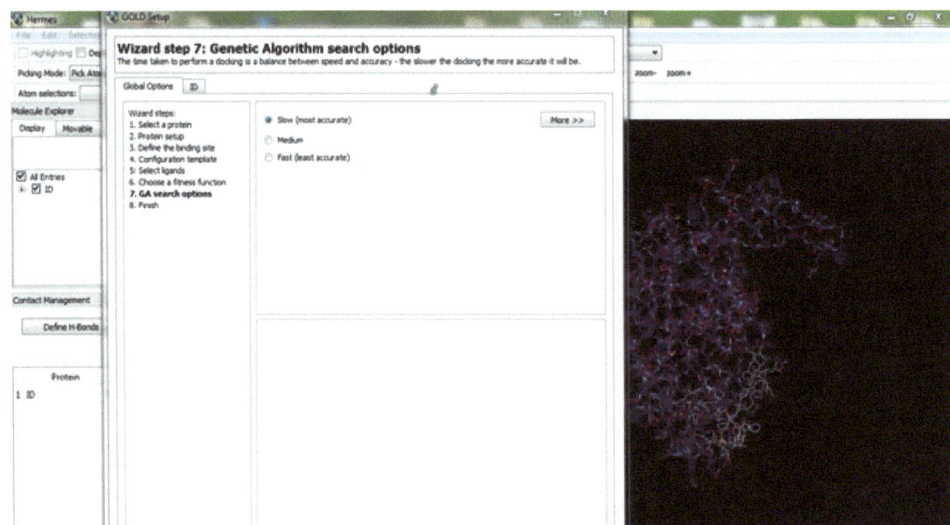

Fig. (137). Parameters selection.

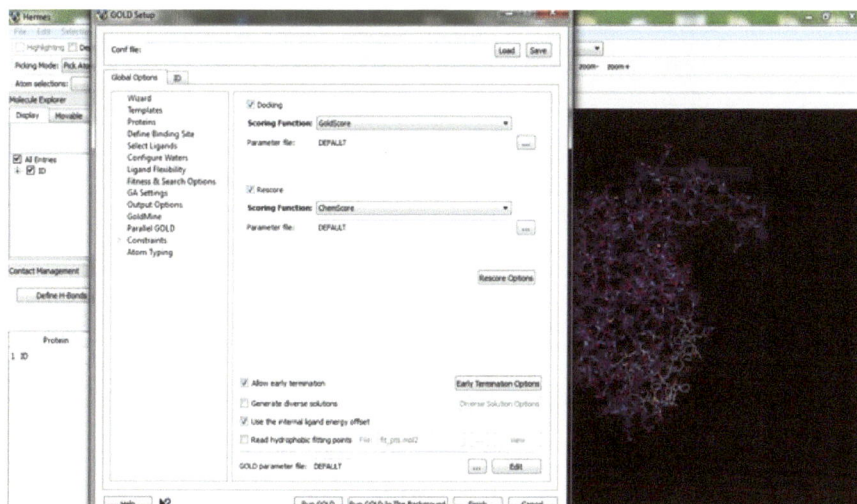

Fig. (138). Selecting scores.

By clicking on the "fitness and search" option from "Global options", "Goldscore" option can be selected and also check the option "Chemscore" for cross verification of the results with other scoring function (Fig. **138**). Click on the "Output options" from "Global options".

The new window will appear. From there select an option "Create output sub-directories for each ligand", "save ligand rank", "save initialized ligand file" (Fig. **139**).

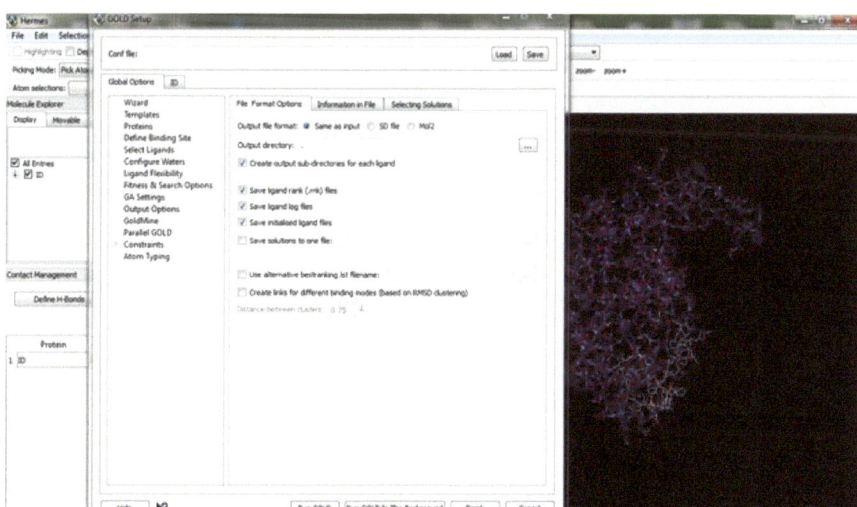

Fig. (139). Save the results.

Remaining "Global options" remained the same and select the "Global options" one by one to just check out the parameters or options which have already selected or can be changed according to the need.

Click on the "Run GOLD" option to run the analyses.

By clicking on the "Run GOLD", and select the option "Messages" for the receiving of message about the outcome either docking start or not (Fig. **140**).

By clicking on the "list of ligand log" option, the log file of ligand will be displayed (Fig. **141**).

The finished GOLD run will appear at the completion of the docking analyses (Fig. **142**).

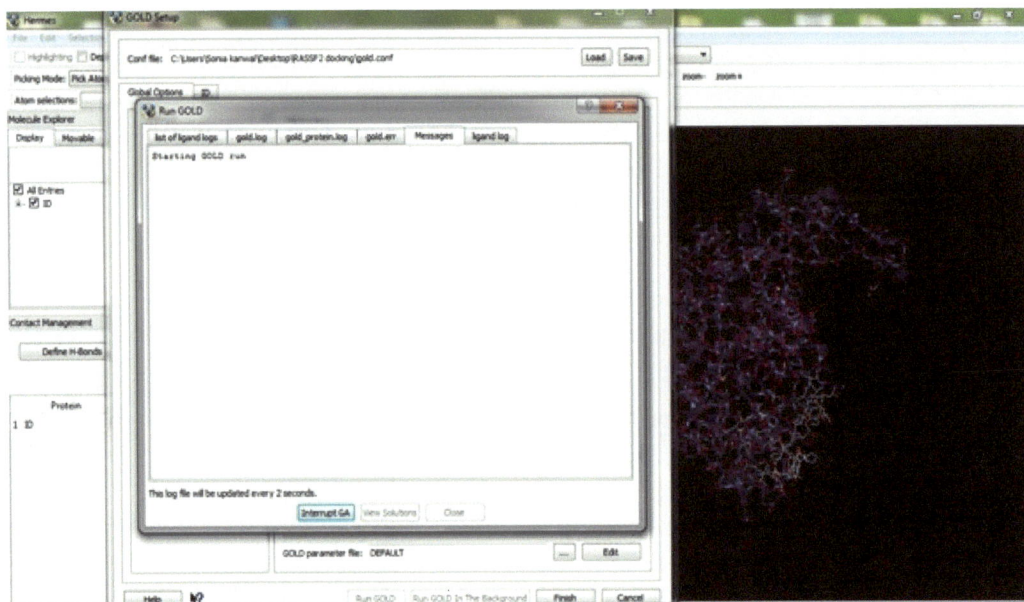

Fig. (140). Run the GOLD.

Results Interpretation

After finishing the "GOLD run", results can be interpreted by the following steps.

There will be following "Docking solutions" with "Goldscore_Fitness" and "Chemscore" on the left side of the Hermes window (Fig. **143**).

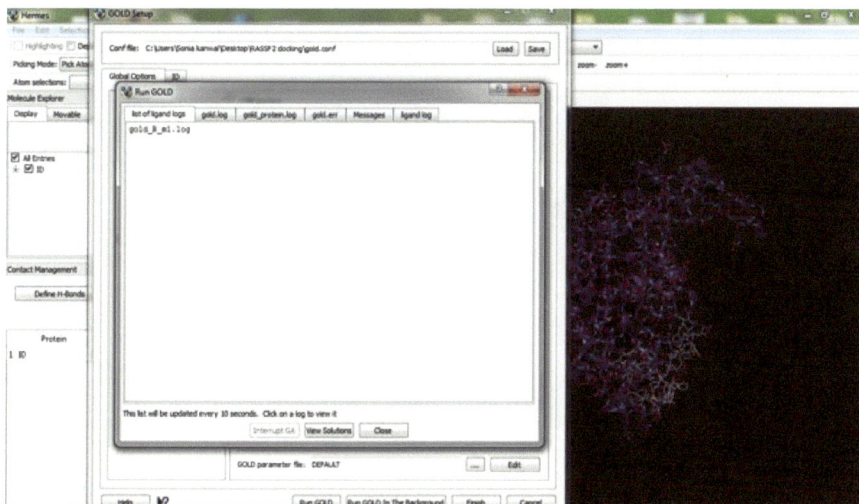

Fig. (141). Results file.

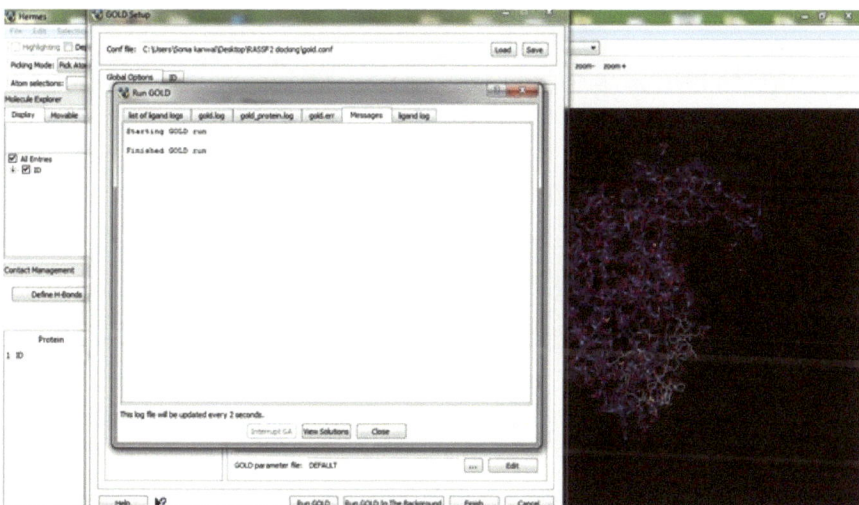

Fig. (142). Job has done.

Click on the "Group by" dropdown list and select the option "Ligand only". Click on "Sort" option and select "Ascending" to sort the results in ascending order (Fig. **144**).

Click on each docking solution and have to save as .pdb and .mol2 files. Click on the first docking solution "Regoravenib.pdb" with respective "Goldscore" and 'Chemscore" fitness values (Fig. **145**). Click on "File" option from the toolbar of "Hermes", click on "Save as" to save the files.

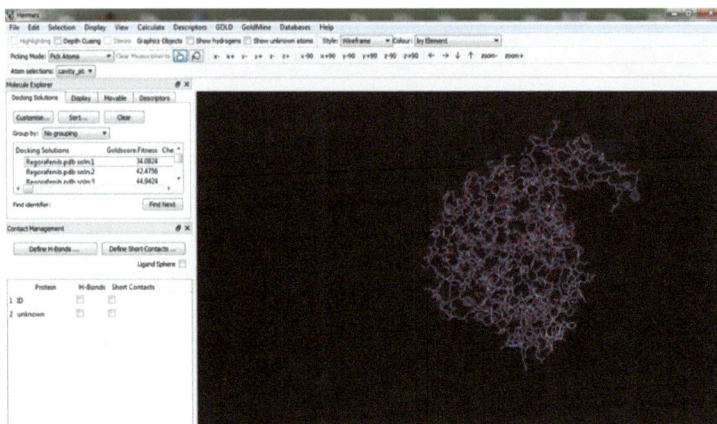

Fig. (143). Generated results.

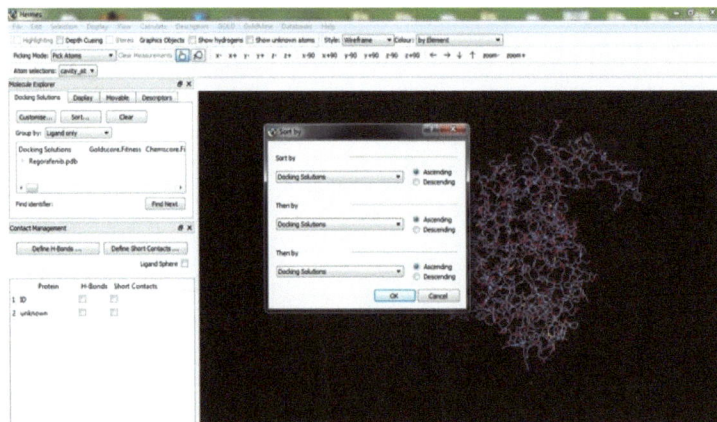

Fig. (144). Results sorting.

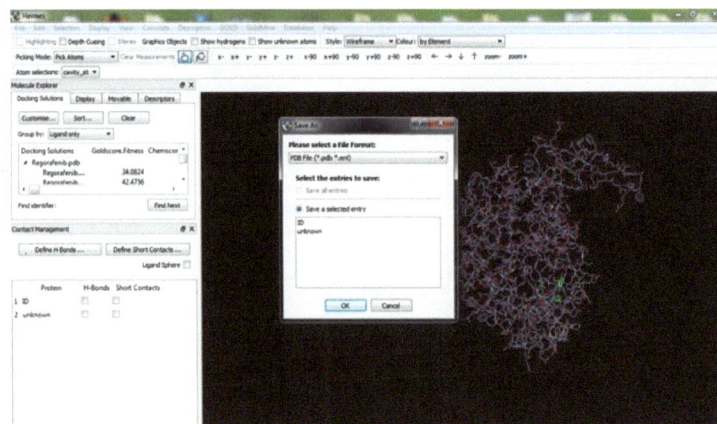

Fig. (145). Selecting of results for analyses.

Select the appropriate folder or drive to save the file in .pdb format (Fig. **146**). The docking solutions are also saved in .mol2 format as shown in Fig. (**147**).

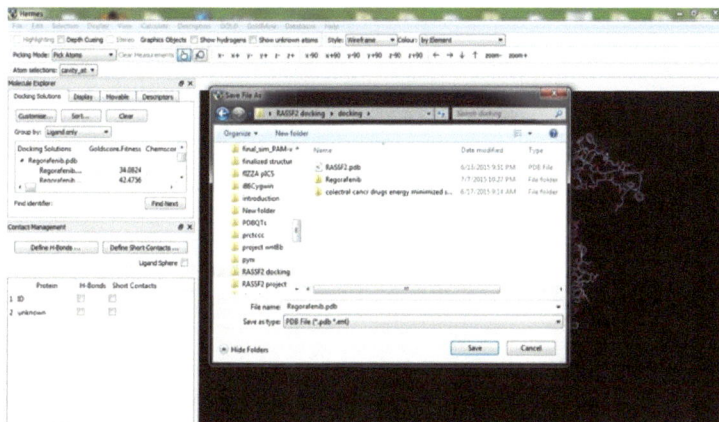

Fig. (146). Saving .pdb files.

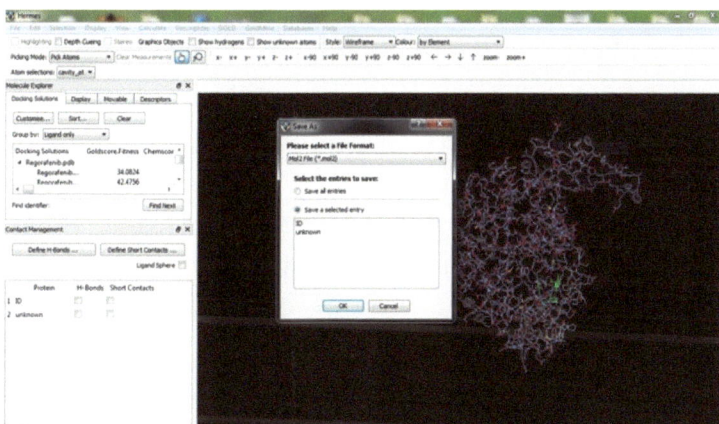

Fig. (147). To save the docking solution as .mol2 file.

Nowadays, bioinformatics is the need of every biologist and every researcher tried to add the computational analyses to their work to bloom up their research work. Usually, the researchers just learn how to run the tool without knowing the sole of the experiment and generate the data. The key component in computational analyses is the input and the interpretation of the results. Tools will generate the data and the researcher has to decide by their literature whether it is gold or the garbage.

LIMITATIONS OF DOCKING TOOLS

AutoDock 4

- Docking is computationally slow through AutoDock 4.
- Docking requires many hours, which makes screening of a large number of molecules impractical.
- It is based on the Genetic Algorithm. The algorithm may fail in case of a large binding cavity of the enzyme.
- Usage of precomputed special grids. A separate grid is built for each type of ligand atoms over the region of space where molecular interaction is expected. On the other hand, grid cells could be made smaller to improve the precision of calculations.

AutoDock Vina

- Vina ignores user-supplied charges.
- Vina avoids imposing artificial restrictions, such as the number of atoms in the input, the number of torsions, the size of the search space and the exhaustiveness of the search.

GOLD

- GOLD is commercialized and needs to be purchased for working.

Molecular docking tools along with details are mentioned in Table **7**.

Table 7. Molecular Docking studies.

Tool Name	Availability	Description	Work Flow	URL	References
AutoDock 4	Freely available desktop tool	It implements genetic algorithm to determine the binding affinities of ligands.	+Receptor +Ligand *Grid, *Poses =Complex =Clusters, Scoring	http://autodock.scripps.edu/	Chang *et al.*, 2010
AutoDock Vina	Freely available desktop tool	It is a fast algorithm to identify interaction of ligands with bio macromolecular targets	+Receptor +Ligand *Add hydrogens *Grid, Conf. file =Binding energies	http://autodock.scripps.edu/	Morris *et al.*, 2009

(Table 7) contd.....

Tool Name	Availability	Description	Work Flow	URL	References
GOLD	Commercially Available	It identifies the binding modes for a large range of test set cases with high confidence rates.	+Input Coordinates *Target site *Scoring Function =Complex =Fitness score	https://www.ccdc.cam.ac.uk/solutions/csd-discovery/components/gold/	Jones *et al.*, 1997

REFERENCES

Bissantz, C., Folkers, G., Rognan, D. (2000). Protein-based virtual screening of chemical databases. 1. Evaluation of different docking/scoring combinations. *J. Med. Chem.,* *43*(25), 4759-4767. [http://dx.doi.org/10.1021/jm001044l] [PMID: 11123984]

Campbell, S.J., Gold, N.D., Jackson, R.M., Westhead, D.R. (2003). Ligand binding: functional site location, similarity and docking. *Curr. Opin. Struct. Biol., 13*(3), 389-395. [http://dx.doi.org/10.1016/S0959-440X(03)00075-7] [PMID: 12831892]

Chang, M.W., Ayeni, C., Breuer, S., Torbett, B.E. (2010). Virtual screening for HIV protease inhibitors: a comparison of AutoDock 4 and Vina. *PLoS One, 5*(8), e11955. [http://dx.doi.org/10.1371/journal.pone.0011955] [PMID: 20694138]

Goodsell, D. S., Morris, G. M., Olson, A. J. (1996). Automated docking of flexible ligands: applications of AutoDock. *J Mol Recognit, 9*(1), 1-5. [http://dx.doi.org/10.1002/(SICI)1099-1352(199601)9:1<1::AID-JMR241>3.0.CO;2-6 [pii]\r10.1002/(SICI)1099-1352(199601)9:1<1::AID-JMR241>3.0.CO;2-6]

Seeliger, D., de Groot, B.L. (2010). Ligand docking and binding site analysis with PyMOL and Autodock/Vina. *J. Comput. Aided Mol. Des., 24*(5), 417-422. [http://dx.doi.org/10.1007/s10822-010-9352-6] [PMID: 20401516]

Sehgal, S. A., Tahir, R. A., Shafique, S., Hassan, M., Rashid, S. (2014). Molecular modeling and docking analysis of CYP1A1 associated with head and neck cancer to explore its binding regions. *J Theor Comput Sci., 5*(112), 2.

Sehgal, S.A., Khattak, N.A., Mir, A. (2013). Structural, phylogenetic and docking studies of D-amino acid oxidase activator (DAOA), a candidate schizophrenia gene. *Theor. Biol. Med. Model., 10*(1), 3. [http://dx.doi.org/10.1186/1742-4682-10-3] [PMID: 23286827]

Trott, O., Olson, A.J. (2010). AutoDock Vina: improving the speed and accuracy of docking with a new scoring function, efficient optimization, and multithreading. *J. Comput. Chem., 31*(2), 455-461. [http://dx.doi.org/10.1002/jcc] [PMID: 19499576]

Verdonk, M.L., Cole, J.C., Hartshorn, M.J., Murray, C.W., Taylor, R.D. (2003). Improved protein-ligand docking using GOLD. *Proteins, 52*(4), 609-623. [http://dx.doi.org/10.1002/prot.10465] [PMID: 12910460]

SUBJECT INDEX

A

Ab initio 1, 22
Acceptor residues interactions 17
Alignment errors 48, 53
Alignment file 46, 48
Alignment sequence 46
Alnfile names 49
Amino acid compositions 12, 13, 15
Amino acids 12, 13, 17, 46, 52, 60, 62, 76, 87
Amino acids sequences 77
Amino acids sequences ID 66
Angles 66, 108 109
  deviated 66
  torsion 108, 109
Atomic compositions 13
Atomic-resolution model 4
Autodock tools 90
Automated web-based modeling tool 54
Automated web modeling 52
Auto model class 49
Automodel class 49
Automodel object 49
Availability description work flow 126

B

Balls and stick model 79
Basic local alignment search tool (BLAST) 5, 46
Binding pocket 90
Bioinformatics 1, 2, 3, 4, 17, 125
Bioinformatics tools 1, 2, 3
Biological data 2
Biological problems 1, 2

C

Chain bond angles 75
Chains, non-redundant protein 60

Chemscore 121, 122, 123
Chi-2 dihedral angle 73
Click on autodock Icon for docking analyses 102
Coiled-coiled structure 12
Comparative modeling 42, 54
Comparative protein structure modeling 44
Computational analyses 1, 125
Computer aided drug designing (CADD) 1, 9
Conf.txt File 113
Critical assessment of structure prediction (CASP) 23
Crystallographic model building 57

D

Docking 6, 90, 91, 94, 112, 117
  blind 112, 117
  flexible 91, 94
  protein ligand 90
  protein-protein 6, 91
Docking analyses 90, 100, 102, 103, 114, 122
  ligand-protein 90
  molecular 90
Docking experiments 91, 107
Docking parameter file 98
Docking parameters 99, 100
Docking solutions 122, 123, 125
Docking tools 126
Domain partition 33
Domains, multiple 25

E

Energy calculations 60, 77
Energy minimization 79, 88
Energy profile 44
Evolutionary biology 1
Extinction coefficient 13, 14

pair-wise 5
Sequence databases 45
Sequences prediction techniques 22
Sequence submission form 35, 38, 43
Sequential number 69, 71
Servers 39, 42, 54
  comparative protein structure modeling 42
  function prediction 39, 54
Silico approaches 2
Software 2, 4, 9, 53, 60, 90, 91
  homology modelling 53
Solving protein crystal structures 27
Stereochemical parameters 68
Stereochemical properties 68, 70, 77
Structural analyses 20, 22, 53, 68
Structural classification of proteins database 25
Structural information 3, 4, 48, 56
Structure analysis 86
Structure assembly simulations 29, 30
Structure editing 88
Structure evaluation tools 75, 76
Structure files 46, 79
Structure prediction, secondary 17
Structure prediction algorithms 23
Structure prediction approach 54
Structure prediction techniques 22
Structure prediction tools 8, 23, 51, 54
  computational 54
Structure visualization tools 88
Surface cleft analysis 18, 20
Swissmodel.expasy.org/interactive 34
SWISS-MODEL homepage 34
SWISS-MODEL site 52

T

Target protein 3, 4, 5, 24, 27, 30, 31, 32, 33, 34, 35, 37, 38, 40, 41, 42, 43, 44, 45, 57, 59, 60, 61, 63, 65, 66, 68, 70, 71, 83, 84, 86, 88, 90, 91, 116

Target protein sequence 23, 25
Target protein submission 39
Target sequence 26, 41, 44, 50, 53
Target-template alignment 5, 49, 53
  suboptimal 53
Template PDB File 47
Template protein 5
  selected 5
Template protein sequence 5
Template protein structure 23
Templates 4, 5, 6, 18, 20, 23, 26, 27, 28, 35, 36, 41, 42, 45, 46, 47, 48, 49, 52, 53, 54, 118, 120
  multiple 42, 54
Template selection 5
Template sequence 4, 26, 44
Templates sequences 5
Template structure 4, 5, 45, 49, 50
Threading template alignment 29
Threading templates used 28, 29
Tools 2, 4, 5, 23, 27, 44, 54, 59, 91, 102
  above-discussed 44
  automated blind docking 91
  based modeling 54
  based structure modeling 44
  cited alignment 5
  command-line 4
  computational 2, 54
  desktop 126
  excellent 42
  model verification 59
  non-commercial 23
  open source 4
  standalone 4
  top-ranked homology modeling/threading approach 27
  uncountable 2
  visualizing 102
  web-based modeling 44
Tools features 88

www.ingramcontent.com/pod-product-compliance
Lightning Source LLC
Chambersburg PA
CBHW041712210326
41598CB00007B/629